T0321225

Chironomidae Larvae

General ecology and Tanypodinae

Chironomidae
Larvae of
the Netherlands
and Adjacent Lowlands

General ecology and Tanypodinae

Henk J. Vallenduuk and Henk K.M. Moller Pillot

KNNV Publishing

CONTENTS

1 INTRODUCTION

Chironomidae larvae play an important role in aquatic ecosystems and are used by water authorities in the assessment of the quality of their water bodies. At present the most urgent problems are difficulties with identifying the species and the lack of information about their biology and ecology. We have tried to simplify the keys and have collected all the available information about the biology and ecology of these larvae from the literature and our own observations.

For normal everyday work it is necessary to identify a species as quickly as possible. To facilitate this the keys make use of as many characters as possible that can be observed with a binocular microscope at 40 x magnification, especially in the first part of the keys. In addition, the keys are accompanied by tables of important characters and some comments to allow the reader to check results by comparing the identified species against related species. Where possible we have described the most useful characters for identifying a species at first glance. Moreover, we have treated the third as well as the fourth instar.

To ensure that the keys did not become over-complicated and difficult to use, they are limited to species living in the Netherlands and the adjacent lowlands. In the text some species of faster running streams in Belgium, Germany and the British Isles are treated briefly. For extensive descriptions and identification of all genera in Europe, see Wiederholm (1983), which also contains a key for the subfamilies (also in Lindegaard, 1997). A catalogue of genera has been published by Ashe (1983) and Palaearctic species are listed by Ashe and Cranston (1990). For important notes about the nomenclature, see Spies and Saether (2004). Larval anatomy has been studied by Thienemann and Zavrel (1916) and Gouin (1959).

This first part of the work covers the subfamily Tanypodinae. Chapter 2 contains information on the general ecology of the whole family which is necessary for the interpretation of samples and species lists.

2 GENERAL ECOLOGY

The general ecology of the Chironomidae is too big a subject to cover fully here. In this chapter we cover only those aspects of particular relevance to the main purpose of this handbook: to describe the ecological characteristics of each species of the Chironomidae occurring in the Netherlands and adjacent lowlands, and to provide information on identifying the larvae. We discuss adult flying and dispersal,[1] the life history of the Chironomidae, sampling methods, fluctuations in numbers and the effects of environmental factors. Much of the literature on the ecology of the Chironomidae has been reviewed by Pinder (1986). For further information we refer to Armitage et al. (1995).

2.1 ADULT FLYING AND DISPERSAL

It is important to have information about movements of chironomids through the air for a number of reasons. For instance, to establish whether a species is able or likely to colonise a specified place at a particular time. The absence of a species can often be attributed to the physical impossibility of reaching an area rather than to the local ecological conditions. In this chapter we do not discuss life cycle in any depth, but the reader should bear in mind that most species do not fly during the whole year. Information about the life cycle of each species is given in Chapter 7.

The movements of bloodsucking Nematocera are relatively well described. Culicidae and Simuliidae are able to migrate over more than a hundred kilometres in one season (see Johnson, 1969) and the females fly in search of hosts to suck blood, especially during summer nights. Female gnats (Ceratopogonidae), which are noticeably smaller, also fly often and far: Brenner et al. (1984) report a mean distance of about a kilometre per night for *Culicoides mohave*. For chironomids it is not useful to cover such distances for food and therefore we cannot assume that it is also normal for them to cover such distances. Data on the move-

ments of chironomids are scarce; females in particular are rarely identified because there are very few keys for females. The following information gives a general impression of adult flying.

Movements for mating and egg-deposition

The main reasons for members of the Chironomidae to fly are to mate and deposit eggs. In most species the males swarm in small or larger groups, but samples of swarms often yield a small number of females (Downes, 1969). The midges most probably swarm not far from the place of emergence so that the females can find the males easy. Many species have specific places for swarming determined by height, distance to the water edge, markers in the landscape, etc., as stated by LeSage and Harrison (1980) for different *Cricotopus* species. Related species often swarm together, as we have observed in *Metriocnemus* and *Smittia*. In one case *Smittia edwardsi* swarmed together with aphids. Cunningham-van Someren (1975) observed many cases of Chironomidae swarming together with Culicidae and/or Chaoboridae. Some species swarm near a high protruding point, such as the top of a tree or a chimney; other species swarm less than one metre above the ground. Small species often swarm nearer the ground and larger species usually swarm above high markers (Gibson, 1945). Some species, for example *Trissocladius brevipalpis* (own observations), do not seem to swarm at all, but mate on the ground or on the water surface. Swarming stops in strong winds.

Species that swarm higher up can be dispersed further, but swarming is not necessarily the only reason for flying. One might expect that males are dispersed further than females because of swarming, but Reiss (1971) observed *Metriocnemus terrester* copulating at a height of 10 m near his house and Peeters (unpublished) found copulating *Metriocnemus fuscipes* and *Smittia edwardsi* at a height of 45 m. It is remarkable that last two species were also found swarming near the ground. Some time can pass between

---

[1] Migration of larvae is treated briefly in section 2.4

mating and oviposition. The females of Tanypodinae deposit their eggs a few days after mating, often on the fifth or sixth day, and die one day later (Koreneva, 1959). Females of each species deposit their egg masses in specific places (see next section).

Other aspects of flying

In the second paragraph we stated that chironomids may not need to fly as far as blood-sucking species because they do not need to fly to find food. However, the supposition that movements of chironomids are mainly related to active flight for mating and egg deposition is doubtful. Johnson (1969) has shown that dispersal is very important for many species of insects, especially for species in unpredictable habitats. McLachlan (1985) and Delettre (1988) found the same in chironomids. Although wing length is not a reliable criterion for predicting habitat characteristics (Delettre, 1988), chironomid species will have more or less innate strategies for dispersal by air, depending on the type of habitat they live in. Delettre et al. (1992) have reported that adults of some terrestrial species were restricted to their specific habitat, whereas other species dispersed widely and also flew above other habitats. McLachlan (1983) found similar differences between two rain-pool dwellers in Africa: *Chironomus imicola* was able to fly for more than an hour and dispersed easily, whereas *Polypedilum vanderplanki* dispersed less, but survived desiccation of rain-pools for more than a year.

Although dispersal by air can be a more or less passive process, Johnson (1969) stresses that most insects actively start migration. He gives daytime catches of Diptera at various heights in the air in Louisiana (from Glick, 1939). Chironomidae were very numerous at 200 feet and still present in fairly high numbers at 5000 feet.

Indications of distances covered

McLachlan (1983) found a radioactively marked female of *Chironomus imicola* at a distance of 847 m from the place of emergence. Moller Pillot (2003) found that egg-depositing females were commonly present at a distance of 450 m from their place of emergence, but that they are probably scarce at a distance of 3000 m. Also, many aquatic chironomids were caught on lawns and other dry places in and around the centre of Tilburg. Some of these belonged to species originating from water bodies at least 500 to 1000 m away, and many of them were large species of aquatic chironomids (which usually swarm high up and are possibly less subject to desiccation). The majority were males (43 males and 18 females), possibly because they remain longer in the air during swarming. The large numbers of chironomids high in the air reported by Johnson (1969) indicate that chironomids can be dispersed over large distances. It is not possible to say what percentage of the females have mated, but the maximum dispersal distance is limited only by desiccation. Among the insects collected by Lempke (1962) on a lightship in the North Sea, 50 km from the coast of Belgium, were species of Bibionidae, Fungivoridae (1 specimen) and Limoniidae. Chironomidae were not collected. It is very likely that such distances are not rare for at least the larger Chironomidae. We have also occasionally found a species 50 km or more from its normal area of distribution. Some species are apparently unable, or rarely able, to colonise the Wadden Sea islands, although suitable water bodies are present.

2.2 THE LIFE HISTORY OF THE CHIRONOMIDAE

In this chapter the development from egg to adult is discussed as far as this is important for understanding the ecology of the species and for identification. Information for identification can also be found in the introduction to the identification of each subfamily. For extensive descriptions see also Branch (1923), Thienemann (1954), Kalugina (1959), Oliver (1971), Aleksevnina and Sokolova (1983) and Armitage et al. (1995). We do not discuss life cycle strategies, but refer the reader to Tauber and Tauber (1976), Tauber, Tauber and Masaki (1986) and to Goddeeris (1983, 1987, 1989). All species of Chironomidae have four larval instars. Most suggested exceptions are usually based on incorrect data, although a few exceptions do exist (McCauley, 1974).

Eggs

The eggs are laid in a gelatinous matrix. The number of eggs in one egg mass varies between species from fewer than ten to several thousand (Munsterhjelm, 1920; Thienemann, 1954; Pankratova, 1977; Nolte, 1993). Some Tanytarsini lay their eggs singly. Females usually produce only one egg mass, but in some cases the eggs are deposited in several groups, for example in some Tanytarsini, in *Pseudochironomus* and in some terrestrial Orthocladiinae (Nolte, 1993). Some other species can also produce a second egg mass, as observed in *Chironomus* (Pinder, 1989). Within a species the number of eggs produced can vary according to environmental factors such as the quality of food and ambient temperature (Palavesam and Muthukrishnan, 1992).

Females often deposit their eggs on a firm substrate (stones, branches, strong plants), but in some cases also directly on the surface of water bodies. Depending on the species, the eggs are laid along the bank, at the edge of the vegetation, etc. (Munsterhjelm, 1920; Aleksevnina and Sokolova, 1983). These sites may also differ between species belonging to the same genus (Strenzke, 1960; Matena, 1990). In large water bodies the majority of eggs are deposited near the banks, but some are thrown off above the deeper parts of the lake (Munsterhjelm, 1920; Wesenberg-Lund, 1943: 514; Aleksevnina and Sokolova, 1983). Many of these egg masses accumulate along shorelines exposed to the wind (Davies, 1976). The egg masses float for a short time before absorbing water and sinking (Sokolova, 1971). In rare cases, they can float for some hours and may be transported downstream (Williams, 1982).

The development of the eggs is temperature-dependent and usually takes a few days. Some species, such as *Tanytarsus holochlorus*, have a diapause in the egg stage (Goddeeris, 1983). Lindegaard and Mortensen (1988: 573, 575) suggest that delayed hatching is also possible (see also Thienemann, 1954: 292). Goddeeris (personal communication) believes that such a strategy does not exist in chironomids. Predation of eggs is common; Hydracarina and fungi can cause substantial losses (Nolte, 1993). The morphological characteristics of egg masses of many species are described by Nolte (1993), who also gives a key for identification.

Larvae

The **first instar larvae** or larvulae are morphologically different from later instars (Kalugina, 1960), but the shape of the head and the position of the eyes is already more or less typical for the different subfamilies. In many Chironomini the median tooth of the mentum is trifid (Kalugina, 1959; Soponis and Russell, 1982); however, some genera have a simple tooth or two median teeth (Soponis and Russell, 1982). In all subfamilies the basal segment of the antenna is very short; only in the Pentaneurini is the antennal ratio (AR) already 3 at this stage (Thienemann and Zavrel, 1916). The AR changes considerably during development because the other segments do not grow much after the first instar. In the Tanypodinae the first instar already has a developed ligula, but (as far as is known) in all species it only has four teeth (Thienemann and Zavrel, 1916: 627; Kalugina, 1959: 93).

The first day after hatching, the larvulae crawl within and around the gelatinous egg mass. On the second day they leave the mass and, at least in stagnant water, lead a planktonic existence for one or two days. This has been found in the Chironominae as well as in the Orthocladiinae and Tanypodinae (Lellák, 1968; Davies, 1976). The larvulae are free swimming and remain on the substrate only for short periods. In all respects they are more mobile than older larvae (Kalugina, 1959): they creep better and can swim much better. All species are probably positively phototactic for one or two days (Davies, 1976) and this period is very important for dispersal and habitat selection. When three or four days old the larvulae begin to settle and some species build tubes (Branch, 1923;

	SEGM. 1	SEGM. 2	SEGM. 3	SEGM. 4	SEGM. 5
instar ii	10	8	5	3	2
instar iii	27	9	6	4	3
instar iv	60	11	8	5	3

Length of antennal segments in *Cricotopus sylvestris* (after Rodova, 1966)

Danks, 1971). The larvulae of *Anatopynia plumipes* begin to creep more frequently on the bottom sediment. Storey (1987) obtained large numbers of *Eukiefferiella ilkleyensis* larvulae by washing water plants (*Ranunculus penicillatus*).

The larvulae begin to feed on small particles, especially fine detritus, in the planktonic stage (Branch, 1923; Davies, 1976; Mackey, 1979). Thienemann and Zavrel (1916) found diatoms and other unicellular algae in the gut of three-day-old Tanypodinae larvulae. *Anatopynia* larvulae at the end of the first instar have been found to attack small animals (own observation). The first instar lasts between 2 and 8 days (Branch, 1923; Lellák, 1968), but in unfavourable circumstances can last more than 14 days (own observations). The larvulae of *Paratendipes albimanus* go into diapause during the summer (Ward and Cummins, 1978). Goddeeris (1983) found no diapause during the first instar in any other species.

Second instar larvae exhibit almost all the **species-specific characters**, both morphological as well as in behaviour and autecology. However, striking morphological differences are visible between the different instars in *Zavreliella marmorata* (Kalugina, 1959: 86) and in *Paratendipes albimanus* (Ward and Cummins, 1978). As a rule, the different instars can be identified by the sizes of the sclerotised parts. In some species, however, there may be some overlap between successive instars (Ladle, Welton and Bass, 1984).

The characters described in the keys, such as the pattern of mental teeth, presence of tubuli, etc., can usually be used to identify instars ii, iii and iv. Head length and head width of third instar larvae are 60% of these lengths in fourth instar larvae, and this 60% ratio also applies to the second and third instar. Some

parts grow more than others during development. For instance, the antennal ratio (AR) increases from the first to the fourth instar because the first segment grows more than the other segments. Rodova (1966) gives the following sizes of the antennal segments of *Cricotopus sylvestris* (see above).

Setae length increases after each moult, but younger instars have relatively longer setae, for example the l_4 of *Cricotopus sylvestris* and the anal setae of *Limnophyes* (own data). The number of setae in a l_4-tuft increases in *Cricotopus sylvestris* after each moult (Rodova, 1966). **Larva size** depends largely on environmental conditions, especially temperature (see par. 2.5), and mature females are larger than males (Oliver, 1971). Laville and Giani (1974) found a clear bimodal curve for the head length of *Ablabesmyia longistyla*, which they ascribe to sexual differentiation.

During the development of the larvae, the sclerotisation of the head and haemoglobin concentration increase and their **colour** gradually becomes more intense. Haemoglobin concentration is temporarily lower during the moult (Kalugina, 1960). In some species, such as *Macropelopia nebulosa*, only the last instar is really red. The colour of sclerotised parts, such as the head capsule, procercus and claws, is usually darker in older larvae. Shortly after moulting, the sclerotised parts have a pale colour and the dark patch on the gula, for example, is not yet visible.

During each instar the teeth of the mentum and mandible can **wear.** , Kalugina (1960) stated this especially after hibernation. The wear is often strongest in the fourth instar and varies between individuals. Certain species or genera have very strong wear (for instance *Stictochironomus* and *Cardiocladius*), depending on their feeding behaviour. Few general characters are known for the **second, third and fourth instar**. Klink

(1983) mentions that *Paratanytarsus* larvae have three chaetulae labrales in the third instar and five in the fourth instar. Just before the end of the third instar the eyes of *Chironomus plumosus* shift to the back of the head (Aleksevnina and Sokolova, 1983). The pupal and adult characters gradually appear during the fourth instar (rarely in the third instar: Goddeeris et al., 2001). The development of the imaginal discs is described by Wülker and Götz (1968) and Goddeeris (1989, 1990). A larva with clearly thickened thorax and abdomen is named a **prepupa**. In principle, prepupae behave as other larvae. In species in which the pupae are enclosed in a protective case (as in many Chironominae and Orthocladiinae), the prepupae build these cases at the end of the larval stage.

In the second instar some larvae of *Chironomus* can still display positive **phototaxis**, which is always absent in the third and fourth instar (Branch, 1923). Markošová (1979) has stated that in *Cricotopus* the second instar larvae in particular colonise new substrates. In *Microtendipes pedellus* the larvae of the first and second instar are positively phototactic (Sokolova, 1971); older larvae display negative phototaxis and they move to the bottom sediment at the beginning of the third instar. In some species negative phototaxis can change to positive phototaxis just before pupation (Kalugina, 1959). Olafsson (1992) stated that in general older larvae are found deeper in the sediment. In one lake, 58.7% of the second instar larvae, 42.1% of the third instar larvae and 20.4% of the fourth instar larvae were found in the top 1 cm. For most species this distribution was independent of the season.

Occasionally, larvae in the third and fourth instar are free swimming. This is found particularly when they are under stress, for example due to a lack of oxygen (Lellák, 1968; Markošová, 1979). Many larvae leave their tube or shelter at night and move to another location (Kalugina, 1959: 100), probably in response to a shortage of food, disturbance by concurrents or predators, or a change in environmental conditions (Moller Pillot, 2003: 57). Disturbance during the day can

lead to an increase in the number of larvae in the water. Unless subjected to heavy disturbance, far fewer fourth instar larvae are carried by water currents (drift) than younger larvae (Moller Pillot, 2003: 60). According to Mason and Bryant (1975) many larvae move from plant surfaces to the bottom sediment in the **autumn**. Some larvae creep deeper into the soil in **winter** (Danks, 1971a: 1883, 1904). However, during the winter typical inhabitants of plant substrates can also be found in large numbers on the surface of plants, even on floating leaves of *Nymphaea* (own observations).

The **duration of the different instars** depends on several factors, especially temperature and food, and is therefore very variable, even within one lake or river. Several forms of stress can hasten or retard their development. For instance, Dettinger-Klemm (2003: 239) found a decreasing development time in *Chironomus dorsalis* and *Polypedilum tritum* as larval densities increased. However, Biever (1971) found a longer development time with increasing density in American species of *Chironomus* and *Tanypus*. Dettinger-Klemm posits that his result for *Polypedilum tritum* is unusual and can be an adaptation to temporary pools; larger larvae in particular increase their rate of development if the pool starts to dry up.

A methodology for **studying the life cycle** is described by Goddeeris (1989), who often observed a regular, continuous growth (Goddeeris, 1983). The duration of the different instars is not the same. A survey of data published by Pankratova (1970, 1977) and Aleksevnina and Sokolova (1983) suggests that, as a rule, each instar lasts longer than the previous one. An example is the development of *Chironomus plumosus* at 18–22 °C described by Pankratova (1977): first instar 5–6 days; second instar, 6–10 days; third instar 6–12 days; fourth instar, 15–19 days. However, these durations may vary widely under the influence of environmental factors and per species (Branch, 1923). **Temperature** has a considerable influence on development time (Oliver, 1971; Balushkina, 1987). An example is given by

Lloyd (1943), who found that a complete cycle of *Limnophyes minimus* lasted 29 days at 20 °C and 260 days at 2 °C. Development times will therefore differ between different climatic regions and between water bodies with different temperature regimes (Borutskij, 1963). **Feeding** is also an important factor. Our own observations of reared *Psectrocladius limbatellus* indicate that after adding food the third instar lasted for a shorter period than the first and second instar. The total development time from egg to pupa lasted from 21 to 33 days depending on the amount of food.

Many species have a **diapause** in the second, third or fourth instar (Grodhaus, 1980; Goddeeris, 1983). The diapause can help the larvae to survive the summer or winter so that their active period falls in the most favourable season. An important function of this can be to synchronise the emergence of adults, which is important for increasing the chances of mating. Some larvae form a **cocoon** for a diapause (Danks, 1971a; Grodhaus, 1980), and Danks reports that the guts of these larvae are always empty. The guts of larvae creeping around during a diapause may be emptied or remain more or less filled. Temperatures near freezing point seem to be responsible for inducing winter cocoons, which have been found mainly in northern latitudes (cf. Danks, 1971a). Goddeeris (1983, 1989) has stated that diapause often starts in a distinct developmental stage, for instance in the summer or autumn, when further development is suddenly stopped at a certain day length. The older larvae continue to grow and will emerge.. After this the whole population is in a distinct developmental stage, forming a single larval cohort. Sometimes individuals are found which have not entered the diapause, especially at higher temperatures.

Pupal stage

The pupal stage is relatively brief, lasting from a few hours to a few days (Oliver, 1971). *Chironomus plumosus* pupae make higher demands on the oxygen content of the water than the larvae (Aleksevnina and Sokolova, 1983). Species living in water with a low oxygen content often have a more developed thoracic horn. The pupae of Tanypodinae are free swimming, but often seek sheltered places on the bottom or between vegetation. Sometimes this can lead to a change of microhabitat, for instance in the genus *Tanypus*, the larvae of which usually live on exposed mud. Many other pupae live in a case, often in the larval tube, or attach themselves to plants or stones. The pupae of many terrestrial Orthocladiinae lie freely on or within the upper layer of the soil or even on moss or grass; other species protect themselves with a gelatinous layer, for example some species in the genus *Metriocnemus*. Pupae of hygropetric species sometimes creep out of the water (own observations).

In aquatic species the **ascent to the surface** is beset with difficulties and dangers and many pupae fall prey to fish and waterfowl. In some cases these pupae are the most important prey for these predators. For surviving of young coots (*Fulica atra*), for example, chironomid pupae must be present in large enough numbers during their first days (Brinkhof, 1995: 72). Species of fast flowing streams often have special adaptations to permit emergence. **Leaving the pupal skin** is also physiologically one of the most critical times in the life cycle. Many reared chironomids appear to get into difficulty just before or during emergence. Bell (1970) reported that if the environment is too acidic the adult characters developed but sometimes no adults were able to leave the pupal skin.

Adults

After emergence, the adults fly to a resting site. Most adults do not feed, but members of many species have been found feeding on honeydew (Downes, 1974). Many adults live for less than one day, but a lifetime of one or two weeks is possible. The males of many species form swarms (see section 2.1) and are attracted to females that enter the swarm. The reproductive isolation of related species is at least partly a result of behavioural differences (Strenzke, 1960). As a rule, the females deposit eggs only once, but they may sometimes lay a second egg mass, probably only after feeding (see Pinder, 1989). Obligatory parthenogenesis is not rare in Orthocladiinae and some Tanytarsini species. Facultative parthenogenesis has been reported in some *Chironomus* species, but in this case the larvae died and only some of them attained the fourth instar (Grodhaus, 1971).

Number of generations

The number of generations is related to the time required for development. Many species are univoltine (one generation a year), others are bi- or trivoltine. In temperate climates, as in the Netherlands, only very few species need two years for development. One example is a part of the *Chironomus anthracinus* populations in deep lakes in Denmark, which have a two year cycle (Jónasson, 1972). Many small species can go through three or more generations a year. If temperature, food and oxygen conditions are optimal (as in rearing conditions), more generations are possible (Lindegaard and Mortensen, 1988: 581). Also in natural conditions the life cycle may vary depending on environmental conditions. This is true for the number of generations as well as for the occurrence of diapause. Goddeeris (1989) points out that very accurate investigations will be necessary to provide a definite answer. For species with more generations a year it may be necessary to take samples every 14 days (at least during the summer) for exact description of the life cycle.

2.3 SAMPLING METHODS

It is not our intention to provide a full review of methods for sampling aquatic invertebrates or flying insects. We discuss only those aspects important for aquatic Chironomidae, and especially the larvae. Brief attention is paid to pupae, exuviae and adults because they are often sampled additionally or as an alternative to sampling larvae.

Sampling time

The best time for sampling and the number of samples that should be taken during the year depend on the life cycle of the species concerned. In running water and in tempo-rary water the fourth instar larvae of many species are more numerous early in spring; in permanent stagnant water many species have a winter diapause and full-grown larvae can be found in greater numbers in summer. Emergence of adults may be more or less synchronised. Shortly after mass emergence only very young larvae are found. Fluctuations in numbers during the seasons are discussed in section 2.4 below. The life cycle of each species is described in the Chapter 7.

2.3.1 SAMPLING OF LARVAE

Sampling method and sample size

The choice of sampling method depends primarily on the aim of the study and how important it is to obtain information on the younger instars. Important points to consider include how much information can be obtained by taking fewer larger samples or a larger number of smaller samples, and how accurate the results should be. Another consideration is weighing up the value of taking samples at several sites against taking more samples at one site, considering the fact that larvae tend to aggregate even in homogeneous environments. More information on choosing between taking a few large samples in one place as opposed to a larger number of small samples can be found in Ringelberg (1976) and Goddeeris (1983). Sometimes there are practical reasons for taking small samples. Hand nets usually become clogged with silt when half a metre has been sampled, making further sampling ineffective. If it is necessary to sort only some of the larvae and estimate the number of remaining larvae, small or tube-living larvae should not be underestimated.

Sampling frequency

The numbers of larvae present can change in a very short time, particularly during the spring and summer (see section 2.4). When studying life cycles it may even be necessary to take samples every 14 days (Goddeeris, 1983). When assessing the trophic state or amount of pollution, the frequency of sampling should be determined mainly by the fluctuations in anthropogenic influences.

Mesh size

The standard mesh size for sampling macroinvertebrates by Dutch water authorities is 500 µm. For specialist investigations a mesh size of 250 µm is recommended (Werkgroep Standaardisatie, 1999). The percentage of living larvae passing the net or sieve is correlated to head capsule width (Storey and Pinder, 1985). These authors found that when washing samples of plants, between 32 and 80% of the second instar larvae of small Orthocladiinae species passed through a 125 µm sieve. Nearly all preserved larvae were retained. Lindegaard and Mortensen (1988) preserved the samples in 4% formalin and used a mesh size of 210 µm. In this case many second instar larvae were retained, including those of rather small species. Goddeeris (1983), who also needed to collect first instar larvae of rather small *Tanytarsus* species, used a mesh size of 70 µm. When sampling drifting larvae in a small lowland brook, Moller Pillot (2003) used a net with a mesh size of 350 x 250 µm to reduce silting up of the net. He lost most second instar larvae, even those of the larger species, such as *Macropelopia* and *Chironomus*, and almost all third instar larvae of the smaller species, such as Tanytarsini, passed through the net.

Special attention to certain groups or species

Larvae living in tubes may remain in their tube while the sample is sorted, which can result in more than 50% of living Tanytarsini being overlooked. Living larvae will leave their tube when the sample is left to dry for five or ten minutes. Mundie (1957: 159) describes another procedure for tube-dwelling larvae in which alcohol and benzene are added several times and shaken very vigorously. Larvae living within stems or leaves can be found only by opening these organs. In the Netherlands, *Stenochironomus* species in particular are rarely found because stems (especially more solid ones, such as *Juncus* stems) are rarely opened. Small larvae are more visible when viewed under transmitted light.

Special methods for different habitats or special purposes

Because sampling small larvae in the conventional manner takes a lot of time, it may be necessary to take special samples for chironomids only. Samples taken for water authority laboratories are usually washed through sieves of different mesh sizes, the main aim being to remove the silt. Many small Chironomidae are usually present on water plants. By washing water plants separately time can be saved and more of these larvae are obtained, if this part of the sample is washed over a sieve with smaller mesh size. Larvae of predatory Chironomidae such as *Kloosia* and *Saetheria* live in sandy sediments in streams. These larvae are very vulnerable and can be found only if the sand is moved very carefully so as not to damage them. *Kloosia* larvae can only be seen under at least 5x magnification.

A significant proportion of Chironomidae live along the banks of streams and stagnant water bodies. Some species are semiaquatic, but species considered as truly aquatic can be underestimated if special attention is not paid to the banks, both below and above the water margin. In streams with coarse sand or gravel, many larvae live up to 1 m deep in the substrate (?terba and Holzer, 1977). In such cases it may be necessary to take samples with a cylinder. In large stagnant water bodies the top 5 cm often contains more than 95% of the larvae (Mundie, 1957). In springs and temporary water bodies and in terrestrial environments it can be better to use an emergence trap in the field or to bring a sample in a mini–emergence trap in the laboratory (see section 3.4). Larvae in running water can also be sampled by using a drift net, in which case silting up is the most important problem. The results are totally different from normal samples, but can give an impression of the larvae drifting in the investigated stretch of the stream (Moller Pillot, 2003).

2.3.2 SAMPLING OF EXUVIAE

Exuviae may be sampled to obtain a more complete survey of all species living in a water body or to identify a higher percentage down to species level. It may be necessary to combine larval and exuviae sampling because of the very large fluctuations in numbers of exuviae and because sampling only exuviae does not reveal where the larvae lived. Exuviae can be sampled without killing the animals, and in many cases the exuviae can be collected easily (Wilson and Wilson, 1984; Wilson and Ruse, 2005). Wilson prefers a hand net; in some cases a larger drift net is necessary. A mesh size of 250 or 500 µm is useful. Klink made a large drift net with a mesh-size of 1 mm for use in large rivers. Information about the method can be found in the publications by Wilson and in Klink and Moller Pillot (1982).

2.3.3 SAMPLING OF ADULTS

Adults may be sampled for different reasons, one reason being that adult males can always be identified down to species level. When discussing the ecology of chironomids in this book we use species names whenever possible because the ecology of related species is usually different. For normal use, adults can be caught with a sweep net. They can also be caught with light traps, but emergence traps provide more accurate results. When adults are not sampled separately it is nevertheless useful to keep adult males, which float on the surface or in spiders webs, in order to make further identification possible if this is necessary.

2.3.4 REARING OF CHIRONOMIDS

Chironomids can be reared in the laboratory in two different ways: rearing separate larvae in dishes, boxes or small vessels, and bringing material into a vessel, cage or mini–emergence trap (photo 1). The first method can be used for studying larval behaviour or to collect larval and pupal skin together with the adult midge to ensure identification. The second method is used for surveys.

Separate rearing
We used small polystyrene boxes with a ventilation cap, placed as shown in photo 2, with the following dimensions: 27 mm wide, 47

mm deep and 65 mm high. Just one larva is placed in each culture box, which contains some material from the collection site and is less than half filled with water from the site. A hole in the cap allows the insertion of a Pasteur pipette, of which the tip is cut off. Most adults will fly towards the light and become trapped in the top of the pipette. The adult can then easily be collected by pinching the neck of the pipette and pouring some alcohol into it. If the adult does not fly into the pipette, some tissue soaked in ethyl acetate can be put in it, in which case the cap has to be closed tightly.

Prepupae give the best results because they do not eat and often the adult emerges in a few days. The larvae are usually not fed. It is not always easy to find the larval skin after the larva has emerged, so it is advisable to add very little material and little food. The cultures should be checked every day. The pupal exuviae always float on the surface of the water. The larval skin can be found in the water, in the larval or pupal case, or sometimes attached to the pupal exuviae.

Rearing for surveys
A larger vessel, cage or mini–emergence trap (photo 1) is used when the aim is to find out what is living in a certain amount of material. Depending on the species and the time of year, it will be some months or more before all the adults have emerged. Bear in mind that an emergence trap will also contain predators and that conditions in the trap will always be different from the natural situation (temperature, oxygen, water movement, etc.). It is often very difficult to find the exuviae from the emerged adults.

2.4 FLUCTUATIONS IN NUMBERS

In most cases the investigator wants to know if the numbers of a species change under the influence of environmental factors. The aim of the investigation may be to determine the effect of anthropogenic influences such as eutrophication and pollution, but food, predation, etc. are often more important. These latter factors are not described here; we refer to the relevant literature, such as Kajak et al. (1968), Jónasson (1972), Kajak (1980), Ten Winkel (1987), Matena (1989) and Moller Pillot (2003). However, fluctuations in numbers can still have other causes. Our aim here is to review these causes.

2.4.1 NUMBERS OF LARVAE

LIFE CYCLE

Numbers of larvae of different instars
The numbers of larvae usually diminish rather quickly during development from the first to the fourth instar. It is not easy to investigate this phenomenon in nature. One reason for this is the shorter lifespan of the younger generations. In a spring population of *Tanytarsus* gr. *holochlorus* Goddeeris (1983: table 59) found a maximum of (only) 2521 larvae in second instar (12 May), 2137 larvae in third instar (25 May) and 2302 larvae in fourth instar (25 May). Without doubt the total number of especially second instar larvae has been much higher than the maximum number. Younger larvae in particular are susceptible to predation by Tanypodinae or other predators, and the presence of many predators can retard the development of a new generation (Matena, 1989).

As a result of the synchronisation of life cycles in most species, fourth instar larvae may be scarce for a short time after the emergence of adults. For the interpretation of data it is important to know when older larvae can be absent. For instance, larvae of *Prodiamesa olivacea* were scarce in many samples taken in May from lowland streams in the province of Noord-Brabant, the Netherlands (Moller Pillot, 1971; 2003). Lindegaard and Mortensen (1988) found hardly any older larvae of *Conchapelopia melanops* in late summer in a Danish lowland stream. The periods of emergence and the presence of fourth instar larvae are described in as much detail as possible in the descriptions of each species in Chapter 7. However, these data usually reflect the situation in the Netherlands and different results may be obtained elsewhere. For example, *Prodiamesa olivacea* emerged in the Danish stream a month later than in the Dutch streams. Also differences are found when water bodies within a region have a different temperature regime. Moreover, many species have a diapause and most are then less mobile, or even live in a cocoon. Such larvae will be found less easily.

Life cycle and methods
As mentioned in section 2.3.1, most investigators of macroinvertebrates in the Netherlands use a hand net with a **mesh size** of 0.5 mm. Nets used to investigate drift should not have a very small mesh size because small meshes will silt up quickly. In such cases the majority of the first and second instar larvae and some of the third instar larvae of small species can pass through the net.

Life cycle and habitat
As stated in section 2.2, younger larvae, especially the first instar, are positively phototactic and are more or less planktonic for some days. The second and third instars of many species tend to live on plants or near the surface, whereas fourth instar larvae are found more on the bottom, a well-known example being Microtendipes (Baz', 1959). The larvae of *Chironomus cristatus* become more negatively phototactic in third instar and are no typical bottom dwellers in second instar (Branch, 1923).

During and after flooding, the number of larvae per square metre in a stream can change dramatically. The fact that younger larvae drift in greater numbers than older larvae is at least partly explained by their positive phototaxis (Moller Pillot, 2003; see also below).

MIGRATION OF LARVAE

Most investigators start from the premise that lower numbers of larvae of a species will be found when environmental conditions are less favourable. However, we should be aware of the fact that larvae migrate for a number of very different reasons. A more extensive review can be found in Davies (1976).

Larvae of *Einfeldia synchrona* move deeper into the mud in summer and in winter

(Danks, 1971; 1971a). Koskenniemi and Sevola (1989) found the same for Chironomus, Glyptotendipes and Polypedilum larvae. From our own observations we know that species living on plants can be numerous in winter at the same sites, but in some cases they live more on the bottom, possibly as a result of preceding frost.

In many species oviposition is often more or less restricted to the litoral zone. In such cases, competition and other factors result in larval migration into the more central parts of a lake. This means that the numbers of larvae in each location change with the season (Kalugina, 1959; Beattie, 1978; Aleksevnina and Bakanov, 1983). According to Koreneva (1959), Tanypodinae always deposit their eggs at the sites where the larvae will live. Romaniszyn (1950) observed an annual migration of Chironomus and Procladius larvae to the profundal zone of a lake in autumn and a return to the litoral zone in early spring (see also Thienemann, 1954: 701–702). The reason for this migration could be to seek water with an optimal temperature and oxygen content. Mundie (1965) also reports vertical migrations of Chironominae larvae, probably in order to escape deoxygenation in the bottom sediment during the summer months.

Larvae can be transported by currents during their life, as Lindegaard and Jónasson (1975) found for Tanytarsus gracilentus in a lake in Iceland. Kalugina (1959) points out that the larvae of many species, including larvae living in tubes, leave their shelter at night to find a better place to live. Moller Pillot (2003) emphasises that nearly every disturbance in flowing water can lead to drift of large numbers of larvae. Shortage of food or shelter often lead to reduced numbers or even the total disappearance of a species in a small stream. Larvae that cannot find a suitable place in a stream may be found in large numbers in drift samples, but rarely in other samples from the stream. For example, Acricotopus lucens, a typical inhabitant of ditches, was apparently only present in drift in the Roodloop stream (ibid., p. 123). A survey of changes in insect–substratum relationships is given by Minshall (1984: 386 et seq.).

2.4.2   Numbers of pupae and exuviae

The numbers of pupae and exuviae are still more dependent on life cycle than numbers of larvae. Mand mass emergence can occur at different times from year to year depending on ambient temperatures. Predation of pupae has been discussed in section 2.2.

Many species of chironomids, including most species that live in tubes (e.g. Chironomus and Tanytarsini), pupate in the same places where the larvae live. However, pupae of Tanypodinae move like culicids, and in many cases they can be found in large numbers among plants and in other places where they are safer from predation. Under normal conditions, exuviae drift no further than 250 m in small streams and up to 2 km in large rivers (Wilson and Ruse, 2005).

2.4.3   Numbers of adults

Quantitative investigations of adults are restricted to emergence studies because flying is affected by the weather and transportation by wind. Fluctuations in the numbers of adults are a result of changes in the numbers of larvae and pupae as described above.

2.5   DIFFERENCES IN BODY SIZE

Larvae, pupae and adults exhibit considerable intraspecific differences in body size. This is not only caused by genetic variation, but also by environmental factors. Much literature exists about seasonal influences (Schlee, 1968; Goddeeris, 1983: 71; Langton, 1991; Kobayashi, 1998). Body size of adults can be 25% smaller in summer than in spring. Temperature is thought to be the main factor in this phenomenon, but food ability and density of larvae also play a role. For instance, Vodopich & Cowell (1984) observed that the mean size of Procladius culiciformis larvae at pupation was significantly larger on a diet of oligochaetes than on a diet of chironomids or zooplankton. McLachlan (1983) has also observed interaction between larval density, feeding and body size. Hildrew et al. (1985) stated that feeding is also dependent on seasonal availability of food. As a rule a combination of factors have to be taken into account when interpreting differences in body size.

2.6 THE EFFECTS OF ABIOTIC ENVIRON-MENTAL FACTORS

The relation between the presence of a species and some abiotic factors is given in the descriptions of the ecology of the species in Chapter 7. This section contains some general points. An important point is that most species have a fairly wide tolerance; the key factor is often the presence and availability of food. The quantity and quality of food and its availability often depend on many environmental conditions, such as temperature, pH, soil and vegetation structure. Vodopich and Cowell (1984) give some examples, which show that the optimal conditions for a species may be at different depths or oxygen contents in different water bodies. They recommend focusing investigations on biological regulators of the prey community and feeding, rather than singular tolerances and preferences of a predatory species.

2.6.1 Temperature

Chironomid larvae are directly influenced by temperature. In most species, growth and development increase as temperature increases (Pinder, 1986). Many authors, including Brundin (1949) and Fittkau (1962), categorise species as cold stenothermous or eurythermous. As well as having a direct influence, temperature also has an indirect influence on larvae, especially through the relationship between temperature and the oxygen content of the water, the activity of harmful chemicals and the quantity and quality of food (Storey, 1987). The same level of organic pollution has quite different effects in summer and in winter, and of course there will be a difference between a shallow pond in open sun and deeper water in shade.
In Chapter 7 we mention the role of temperature only when necessary.

2.6.2 Permanence

In this book a water body is called dry when the bottom is dry. In such cases there will always be a low degree of humidity remaining, but aquatic larvae without special adaptations will die. Nearly all species can occur in temporary water bodies when they are filled with water. When we mention occurrence in temporary water we mean only pools, brooks and ditches that dry out completely in summer (including the upper courses). Species may be found in these places in spring if in the preceding autumn the females were able to lay eggs there. For most species this means that the pool or ditch must have contained water during the flying period of the species that autumn. If water is still present later in spring a second generation can be found.

2.6.3 pH

Emergence of adult chironomids is inhibited when the pH is too low (Bell, 1970). If conditions become too acidic (very low pH) the larvae will die. The quality and quantity of food can also be affected by pH, especially via its effect on decomposition by bacteria (Egglishaw, 1968; Chamier, 1987). Such indirect influences can explain why a species is rarely found in water with a very low or very high pH. A particular problem when interpreting data is that the pH of the organic material in bottom sediments is usually not as extreme as in the water. Goddeeris (1983: 89) found that the pH of the water could be approximately 9, whereas the silt layer was almost neutral.

2.6.4 Oxygen

Oxygen is one of the most important factors influencing the distribution and development of chironomid larvae. As a consequence, many species are not able to live outside cold and/or running water. A reduction in the oxygen content induces larvae at the bottom of streams and lakes to leave their burrows and start migrating (Heinis et al., 1989). For most aquatic animals the frequency and length of periods of low oxygen content are the critical factors. These usually occur during the night. The table on saprobity and oxygen content (Table 2 on p. 130) includes categories for the duration of low levels of oxygen content.

The larvae of many chironomid species possess haemoglobin, which is thought to facilitate oxygen transport and storage (Heinis, 1993). Larvae are also able to switch from aerobic to anaerobic metabolism by glycogenolysis. Heinis found that small larvae have less resistance to anoxia than larger larvae, possibly because they have a higher metabolic rate and exhaust their glycogen supply more rapidly. Heinis and Swain (1986) proved that the larvae respond to reduced oxygen concentrations by investing more and more time in ventilatory movement. Below a certain level they also have to stop eating. Int Panis et al. (1995) investigated the influence of oxygen microstratification in the bottom sediment. Larvae living at the surface of the sediment, at a depth of less than 1.5 cm, are correlated with the oxygen concentration in the water column. Species living at about 2 mm depth, such as *Polypedilum nubeculosum*, are dependent on the oxygen concentration at the sediment–water interface. Fourth instar larvae of *Stictochironomus* are able to live somewhat deeper in the sediment and are less dependent on the oxygen concentration at the surface, possibly by migrating to the sediment surface to breathe and store oxygen by binding with haemoglobin.

2.6.5 Trophic conditions

Large quantities of data are available on the presence of species in relation to the phosphate and nitrogen content of the water. However, there is most probably no true direct relation between these nutrients and the presence of chironomid larvae. Such correlations are therefore only valid for the specific environmental conditions under which these quantities were measured. For example, in a dark wood high nutrient contents do not lead to high levels of production or decomposition. Lindegaard et al. (1975) found *Trissopelopia longimanus*, a typical clean-water species, in a spring with a rela-

tively high phosphate and very high nitrate load. Nevertheless, in many cases the published data can give an impression of the needs (not the tolerance) of a species.

2.6.6 Saprobity

When interpreting saprobity we follow Sladecek (1973: 28) and take it to mean the amount and intensity of decomposition of organic matter. The amount of oxygen is related to saprobity, but can vary between different water types with the same saprobity. The suitability of a habitat for chironomid larvae is determined by the oxygen content, the availability of organic food and the toxicity of the products of decomposition. Because more oxygen is available near the water surface, some species will concentrate there when the water is polluted. Such species therefore have a relatively high tolerance for water pollution.

2.6.7 Trace metals and other chemicals

Nothing is known about the influence of harmful chemicals on most species of chironomids. According to Timmermans (1991) the accumulation of trace metals results in mortality, reduced growth and delayed development of the larvae. Heinis et al. (1990) showed that during exposure of *Glyptotendipes pallens* to high cadmium concentrations food uptake ceased completely.

2.6.8 Salinity

In brackish environments salinity is the dominant environmental factor determining the presence of chironomids (Parma and Krebs, 1977). Data on tolerance of salinity in the Netherlands can be found in Krebs (1981, 1984, 1990) and Steenbergen (1993). In the Baltic states several species have been found in waters with a much higher chlorine content than in the Netherlands (Tõlp, 1971). The most probable reasons for this are that the salinity of Dutch waters fluctuates more and that as a result of upward seepage the salinity of the silt in Dutch waters can be significantly higher than in the water above (Parma and Krebs, 1977: fig. 2). Parma and Krebs (1977) concluded that the chloride concentration in the free water layer accurately reflects the environment to which tube building larvae are exposed. Conditions may be less favourable for species that exhibit less ventilation movement, such as Tanypodinae.

G

3 LIST OF TRIBUS, GENERA AND SPECIES OF THE TANYPODINAE

| * | not entered in the keys |
| ** | only entered in the keys as total genus |

Anatopyniini	*Anatopynia plumipes* (Fries, 1823)
Coelotanypodini	*Clinotanypus nervosus* (Meigen, 1818)
Macropelopiini	*Apsectrotanypus trifacipennis* (Zetterstedt, 1838)
	Macropelopia adaucta Kieffer in Thienemann & Kieffer, 1916
	Macropelopia nebulosa (Meigen, 1804)
	Macropelopia notata (Meigen, 1818)
	Psectrotanypus varius (Fabricius, 1787)
Procladiini	*Procladius (Holotanypus) choreus* (Meigen, 1804)**
	Procladius (Holotanypus) crassinervis (Zetterstedt, 1838)**
	Procladius (Holotanypus) culiciformis (Linné, 1767)**
	Procladius (Holotanypus) ferrugineus (Kieffer, 1918)**
	Procladius (Holotanypus) sagittalis(Kieffer, 1909)**
	Procladius (Holotanypus) signatus (Zetterstedt, 1850)**
	Procladius (Psilotanypus) flavifrons Edwards, 1929*
	Procladius (Psilotanypus) imicola Kieffer, 1922**
	Procladius (Psilotanypus) lugens Kieffer, 1915**
	Procladius (Psilotanypus) rufovittatus (Van der Wulp, 1874)**
	Tanypus kraatzi (Kieffer, 1912)
	Tanypus punctipennis Meigen, 1818
	Tanypus vilipennis (Kieffer, 1918)
Natarsiini	*Natarsia nugax* (Walker, 1856)**
	Natarsia punctata (Meigen, 1804)**
Pentaneurini	*Ablabesmyia longistyla* Fittkau, 1962
	Ablabesmyia monilis (Linnaeus, 1758)
	Ablabesmyia phatta (Egger, 1863)
	Arctopelopia barbitarsis (Zetterstedt, 1850)
	Arctopelopia griseipennis (Van der Wulp, 1858)*
	Arctopelopia melanosoma (Goetghebuer, 1933)*
	Conchapelopia hittmairorum Michiels & Spies, 2002*
	Conchapelopia melanops (Meigen, 1818)
	Conchapelopia pallidula (Mcigen, 1818)*

Guttipelopia guttipennis (Van der Wulp, 1859)

Krenopelopia binotata (Wiedemann, 1817)**
Krenopelopia nigropunctata (Staeger, 1839)**

Labrundinia longipalpis (Goetghebuer, 1921)

Monopelopia tenuicalcar (Kieffer, 1918)

Nilotanypus dubius (Meigen, 1804)

Paramerina cingulata (Walker, 1856)
Paramerina divisa (Walker, 1856)*

Rheopelopia maculipennis (Zetterstedt, 1848)**
Rheopelopia ornata (Meigen, 1838)**

Schineriella schineri (Strobl, 1880)

Telmatopelopia nemorum (Goetghebuer, 1921)

Telopelopia fascigera (Verneaux, 1970)*

Thienemannimyia carnea (Fabricius, 1805)**
Thienemannimyia festiva (Fittkau, 1962)**
Thienemannimyia northumbrica (Edwards, 1929)**
Thienemannimyia pseudocarnea (Murray, 1976)**

Trissopelopia longimanus (Staeger, 1839)

Xenopelopia falcigera (Kieffer, 1911)**
Xenopelopia nigricans (Goetghebuer, 1927)**

Zavrelimyia barbatipes (Kieffer, 1911)**
Zavrelimyia hirtimana (Kieffer, 1918)**
Zavrelimyia melanura (Meigen, 1804)**
Zavrelimyia nubila (Meigen, 1830)**
Zavrelimyia signatipennis (Kieffer, 1924)**

4 INTRODUCTION TO THE KEYS OF LARVAE OF TANYPODINAE

COLLECTING TANYPODINAE LARVAE

See section 2.3.1 for general information about sampling larvae. When **sorting** the larvae the Tanypodinae can already easily be distinguished from other subfamilies. Their heads are relatively large – those of the Pentaneurini also rather long – and most species move backwards when touched. The posterior parapods are spread outwards. All larvae are free-living. To **preserve** them we recommend putting all the collected larvae into a dish with a little water and pouring hot water (at least 60 ° C) over them. When treated this way the larvae stretch out the claws of the posterior parapods so that these can be observed easily when identifying the species. Another option is to put the living larvae into Keyl's solution (3 parts 96% ethanol and 1 part glacial acetic acid). Again, the larvae open out their claws, and this solution makes the head transparent. Larvae can be stored in this solution for a long time.

RECOGNISING THE SUBFAMILY TANYPODINAE

In Tanypodinae the antenna is never on a pedestal, but is retractile and can be drawn into a tube inside the head. The larvae of the Tanypodinae have only one eye-spot (only *Guttipelopia* larvae usually have a divided eye-spot) (photo 3). The central part of the mentum is weak, without teeth, and the larvae have a movable ligula. Lindegaard (1997) gives a useful key to the subfamilies, and a complete diagnosis is given by Fittkau and Roback (1983).

IDENTIFYING LARVAE

For examination we recommend placing the larvae in a petri dish with a solution of 70% alcohol. Many characters can be seen at 40 x magnification. Setae can be seen best with diascopic illumination; some other characters are easier to see with illumination from above. For examining more details we recommend putting a larva on a slide in a drop of alcohol or in a mixture of alcohol and lactic acid (3:1). In a solution of lactic acid the head will become transparent and the setae can be observed more easily. Do not use KOH because setae will then become invisible within a short time. In many cases the larva has to be studied in an exact dorsoventral position. Some larvae are easy to put on a slide in this position, but often it will be necessary to roll the larva. To do this, place the larva next to a pile of two or three coverslips in a drip of alcohol and covered by one coverslip (photo 4). The larva can be rolled by moving the coverslip to the left or right with one fingertip, and the larva will not be flattened.

RECOGNISING DIFFERENT INSTARS

There exists no general rule to distinguish different instars of Tanypodinae larvae. The pupal and adult characters gradually appear during the fourth instar. In Chapter 5.1 there is a key usable for larvae in the third and fourth instar and in Chapter 5.4 we give a key usable only for larvae at the end of the fourth instar, of which the thoracic horn already has been developed.

MEASURING

The length and width of the head must be measured before putting a coverslip over the larva to avoid flattening the head at all. The labrum always presents a problem when measuring the length of the head: sometimes the labrum reaches far forwards, at other times it is completely bent downwards. The sizes given are therefore not exact. Measurements of various other features are best made when the larva is put under a coverslip. The ideal horizontal position of the feature, which must be measured, can be obtained by moving the coverslip.

SPECIAL CHARACTERS

Setae
Setae growing from under the cuticle and a ringwall can be seen where they pass through the cuticle, although this wall is not always obvious. Our data on total number of setae may include underestimates and so the ranges we give may be incorrect. The exact length of the seta on the base of the posterior parapod cannot always be determined because it is often partly broken off. Cephalic setae are also difficult to examine and so we did not use them. Descriptions of this character can be found in Rieradevall and Brooks (2001).

Occipital margin
The occipital margin (a rim at the back of the head) runs around the head and differs in many species on the ventral and dorsal sides. This feature is easily to examine using a binocular. The margin must be in a horizontally position otherwise the shape looks different, but it is usually already in this position. The shape and pigmentation of the occipital margin are good characters for identifying the genera and species.

Tentorial line and pit
The tentorial line can be seen easily. However, its shape looks different when viewed through a microscope and when using a binocular with striking illumination. In our key the shape of the line is not relevant. The tentorial pit is in fact a point where a band inside the head touches the cuticle, what is best seen with a microscope. Sometimes this inside band seems to be located on the surface of the cuticle.

Pigmentation
Dark pigments are absent the first day after moulting and become gradually less strong after storing for one or more years.

USING THE KEYS, MATRIX TABLES AND COMMENTS
The keys refer only to the most important characters. The matrix-tables provide an overview of many other characters. Some species can be identified more quickly using the matrix tables rather than the key. If problems are encountered or identification is doubtful, we recommend consulting the matrix tables and comments.

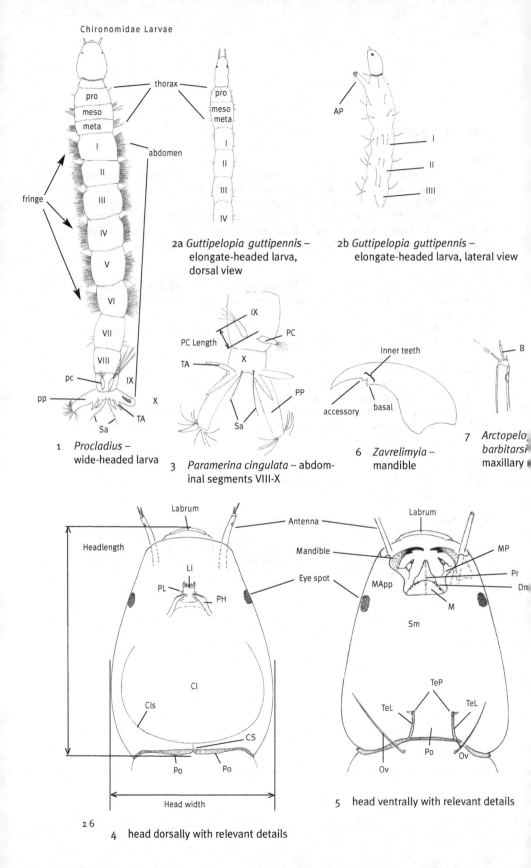

Chironomidae Larvae

thorax

pro
meso
meta

abdomen

fringe

pro
meso
meta

I
II
III
IV

2a *Guttipelopia guttipennis* –
elongate-headed larva,
dorsal view

AP

I
II
IIII

2b *Guttipelopia guttipennis* –
elongate-headed larva, lateral view

pro
meso
meta
I
II
III
IV
V
VI
VII
VIII
pc   IX
pp   X
         TA
Sa

1   *Procladius* –
wide-headed larva

IX
PC Length
TA   X
                PC
                PP
         Sa

3   *Paramerina cingulata* – abdominal segments VIII-X

Inner teeth

accessory   basal

6   *Zavrelimyia* –
mandible

B

7   *Arctopelo...
      barbitars...
      maxillary...*

Headlength

Labrum
Li
PL   PH

Cls

Cl

CS

Po   Po

Head width

Antenna

Mandible

Eye spot

Labrum

MApp

M

Sm

TeP
TeL   TeL
TeL   Po   Ov
   Ov

MP
Pr
Dm

5   head ventrally with relevant details

2 6

4   head dorsally with relevant details

GLOSSARY

CHARACTER	FIG	ABBR	PHOTO	DESCRIPTION
abdominal segments	I			segments I-X, sometimes difficult to distinguish
anal tubules	1,3	TA	11	tubules placed above and between posterior parapods
antenna	4,5,8		23	in Tanypodinae the antenna is retractile
anterior parapod	2b	AP	7	false ventral foot on thorax carrying apical claws
b seta	7	B	36	one seta of a group of setae apically of the maxillary palp
clypeal suture	4	ClS		suture of the clypeus
clypeus	4	Cl		in Tanypodinae the clypeus is covering the entire head
coronal suture	4	CS		suture between clypeal suture and postoccipital margin
dorsomental teeth	5	Dm teeth		teeth on the rim at both sides of the dorsomentum
eye spot	4,5			
fringe (of lateral setae)	I	fringe		lateral setae of abdominal segments; in elongate-headed species not always exactly placed laterally
head length	4			length from postoccital margin to the anterior margin dorsally. Sometimes the labrum not included
head width	4			width of head taken where the head is the widest
index capitis		IC		head width divided by head length
inner teeth of mandible	6			teeth consisting of a basal and accessory tooth in Tanypodinae
labrum	4,5			
ligula	4	Li	43	sclerotized and toothed plate, only in Tanypodinae
mandible	5,(6)			
maxillary palp	5	MP	36	palp placed ventrolaterally on maxilla
mentum	5	M		double-walled medioventral plate consisting of ventromentum and dorsomentum
mentum appendage	5	MApp		
Ov setae	5	Ov	33	setae at anterior margins of all segments ventrally
paraligula	4	Pl	46	deeply serrated plate at both sides of the ligula
pecten hypopharyngis	4	PH	42	one row of teeth to each side on the hypopharynx, crossing the ligula
posterior parapod	1,3	PP		false ventral foot on segment X carrying apical claws
postoccipital margin	4,5	Po		sclerotized rim of head capsule
procercus	1,3	Pc	13	dorsally sclerotized tubercle on segment IX
procercus length	3	Pc length		length of sclerite plate
pseudoradula	5	Pr	38	longitudinal band with granulate surface in the middle of the mentum appendage
submentum	5	Sm		ventral area of head capsule posterior to mentum
supraanal setae	1,3	Sa		setae dorsally on segment X between anal tubules
tentorial line	5	TeL		
tentorial line interdistance	5	int-TeL		distance between both tentorial lines
tentorial pit	5	TeP		
thorax	1,2a, 2b	Th		second part of the body; consisitng of pro-, meso- and meta-thorax (segments often difficult to distinguish)

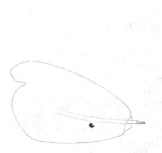

8 *Clinotanypus nervosus* – head lateral view

meso

9 *Psectrotanypus varius* – thorax ventrally (figure: B. Goddeeris)

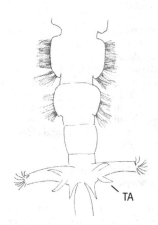

TA

10 *Tanypus kraatzi* – abdominal segments VI-X ventrally

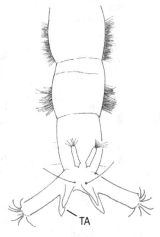

TA

11 *Psectrotanypus varius* – abdominal segments VI-X dorsally

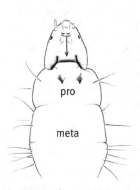

pro

meta

12 *Tanypus kraatzi* – thorax ventrally (figure: B. Goddeeris)

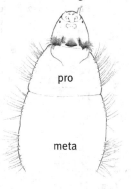

pro

meta

13 *Tanypus punctipennis* – thorax ventrally (figure: B. Goddeeris)

14 *Tanypus punctipennis* head ventrally

5   KEYS

IMPORTANT REMARKS FOR USING THE KEY.

Characters
For additional characters see Comments. An overview of characters is given in the matrix tables. Some species are difficult to identify, especially in the group of larvae without a lateral fringe of setae. We therefore recommend using the characters and the head length given in the matrix tables. The pigmentation of antenna and claws is usually reliable for instar iv larvae.

Setation
Setae characters used in this key are visible using binocular (maximum 100x magnification). Abdominal setae are best seen with diascopic illumination. More small or thin setae will be visible when examined under a laboratory microscope. Because setae can be broken off we advise examining several segments.

Ligula
Note that the ligula can turn over so that it can point forwardly or backwardly. Because of this it is sometimes difficult to observe the margin of the ligula teeth.

Measurements
Considerable deviations can occasionally occur.

5.1   KEY TO THE INSTAR iii AND iv LARVAE OF TANYPODINAE

1 a	Head wide: IC ≥0.7 (fig 1). Abdominal segments I-VI with a lateral fringe of setae (best seen when the larva is placed dorsoventrally and with diascopic illumination) (fig 1 – fringe). Head ventrally with tentorial pit and/or tentorial line (fig 5 – TeP, TeL, photo 5). Anal tubules triangulate (fig 1 – TA)	**2**
b	Head elongate: IC ≤0.7 (fig 2). Abdominal segments without a fringe of setae (fig 2). Head usually without tentorial pit or line. In some species a small tentorial pit is present. Some species have a weak and unpigmented tentorial line, but the line never reaches the postoccipital margin (as in fig 39, photo 6 & 26). Anal tubules usually slender (fig 3 – TA)	**14**

2 a	Head in lateral view wedge-shaped (fig 8). Postoccipital margin with deep incision (fig 8). Mesothorax with 30-60 (or more) lateral setae. Antenna very long, about 3/4 of head length	***Clinotanypus nervosus***
b	Head roundish. Postoccipital margin without deep incision. Mesothorax maximum 21 lateral setae (fig 9 - meso). Antenna shorter than 1/2 head length	**3**

3 a	Two groups of 3 anal tubules (fig 10 - TA)	**4**
b	Two groups of 2 anal tubules (fig 11 - TA)	**5**

4 a	Lateral setae prothorax ≤5, metathorax ≤10 (fig 12 – pro, meta). No pigmentation between the tentorial lines (fig 12 - ↑). Fringe of lateral setae on thorax and abdominal segments almost in one line	***Tanypus kraatzi***
b	Lateral setae prothorax ≥9, metathorax in instar iv ≥20 and instar iii probably only 7-10 setae (fig 13 – pro, meta). Triangular pigmentation spot between the tentorial lines a (fig 14 - ↑). Fringe of lateral setae on thorax and abdominal segments in a broad band	***Tanypus punctipennis***

15 *Procladius* - abdominal segments VI-X dorsally

16 *Anatopynia plumipes* – head ventrally

17 *Procladius* – postoccip region ventrally

18 *Psectrotanypus varius* – thorax ventrally (figure: B. Goddeeris)

19 *Psectrotanypus varius* – postoccipital region ventrally

20 *Psectrotanypus varius* ligula

21 *Psectrotanypus varius* – mandible

22 *Tanypus vilipennis* – thorax ventrally (flgure: B. Goddeeris)

23 *Tanypus vilipennis* – pc occipital region ventral

5 a   Ligula black (best seen ventrally or by pressing the headcapsule dorsally). Segment VII with up to three small lateral setae, much shorter than those on the other segments (fig 15 - VII)   ***Procladius***

b   Ligula yellow, orange or brown. Segment VII with a lateral fringe of more than 3 setae, almost as long as those on other segments (as in fig 11)   **6**

6 a   Tentorial line produces a second sclerite band ventrolaterally as broad as the postoccipital margin (fig 16 - ↑, photo 8). Lateral fringe of setae on segment VII with ≥60 setae. Large species. Body length instar iv: 12.5-19 mm   ***Anatopynia plumipes***

b   Tentorial lines do not produce a second sclerite band; ventrolaterally only the postoccipital margin. In some species there is a second line, but never strongly pigmented (fig 17 - ↑). Lateral fringe of setae on segment VII with maximum 40 setae   **7**

7 a   Head pale (photo 10). In instar iv prothorax with 6-12 lateral setae; instar iii with 3-5 lateral setae (fig 18 – pro, photo 10). Tentorial line not pigmented, sometimes it seems to be absent, but tentorial pit obvious (fig 19). Ligula with 4 equally long teeth (fig 20). Mandible with (4-)6 inner teeth (fig 21), in instar iii (2-)4 inner teeth
    Psectrotanypus varius

b   Head pale, orange or brown. Prothorax with one or two lateral setae. Tentorial line and pit not with this combination except in Macropelopia instar iii larvae. Ligula with 5 teeth, median tooth shorter than adjacent teeth. Mandible with 1 basal tooth and 1 accessory tooth (fig 6)   **8**

8 a   Lateral setae mesothorax maximum 3, metathorax maximum 6 (fig 22 – meso, meta). Head pale. Tentorial line narrow and relatively very short, approximately $^1/_3$ interdistance tentorial lines (fig 23, photo 9). Procercus sclerite pale (photo 11). Two small claws of the posterior parapod with bulbous base (fig 24). Ligula teeth equally long (fig 25, photo 12)   ***Tanypus vilipennis***

K

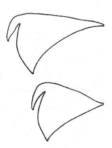

24 *Tanypus vilipennis* –
    claws of posterior parapod
    (figure: B. Goddeeris)

25 *Tanypus vilipennis* –
    ligula and paraligula

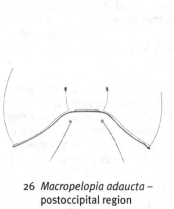

26 *Macropelopia adaucta –* postoccipital region ventrally

27 *Macropelopia adaucta –* ligula

28 *Apsectrotanypus trifascipennis –* head ventrally

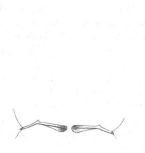

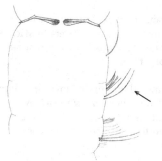

29 *Macropelopia adaucta –* postoccipital margin dorsally

30 *Macropelopia adaucta –* thorax and postoccipital margin dorsally

31 *Macropelopia nebulosa* thorax dorsally

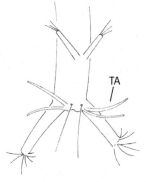

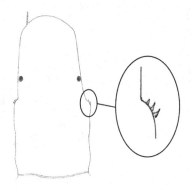

32 *Labrundinia –* abdominal segments VIII-X dorsally

33 *Labrundinia –* ligula

34 *Labrundinia –* head dorsally

34a spines

b More lateral setae on meso- and metathorax. Head brown or orange. Tentorial line relatively long, $\geq^{1}/_{2}$ interdistance tentorial lines (fig 26). Procercus sclerite brownish pigmented (photo 13). No claw of the posterior parapod with bulbous base. Ligula middle tooth smallest (fig 27) **9**

9a Head length 750-1260 µm. Head width at least 510 µm (Instar iv larvae) **10**
b Head length 430-750 µm. Head width maximum 550 µm (Instar iii larvae) **13**

10 a Head capsule brown with pale area around the eye spot (photo 3a). Head ventrally with an obvious pale and crenated longitudinal band (fig 28). Tentorial line broad and strongly pigmented, tentorial pit seems to be absent (fig 28 - TeL). Postoccipital margin ventrally and dorsally uniformly pigmented; dorsally the same width as lateroventrally (fig 28 - ↑Po) *Apsectrotanypus trifascipennis*
b Head capsule pale or orange, sometimes with a brown moisture (bloom) (photo 3b). Pale band inconspicuous. Tentorial line not or weakly pigmented and tentorial pit obvious (fig 26). Postoccipital margin ventrally and dorsally not pigmented; dorsally broadened with a dark median line and in the middle some pigmentation (fig 29) **11**

11 a Mesothorax with a short lateral fringe. Fringe only anteriorly of two robust, somewhat dorsally placed, setae (fig 30 - ↑). Lateral setae on segment VII maximum 3-13 (mean 7.1). Procercus length 320 – 380 µm *Macropelopia adaucta*

Note
M. notata sometimes without setae posteriorly of the two robust setae, but procercus length 250-280 µm.

b Lateral fringe on mesothorax longer. Two robust, somewhat dorsally placed setae, approximately halfway the fringe (fig 31 - ↑). Latereal setae on segment VII maximum 3-32. Procercus length 250 – 320 µm **12**

12 a Lateral setae on segment VII maximum 11-32 (mean 18). Number of setae posteriorly of the two robust setae: 3-15. Thin yellow band along the coronal suture and the posterior part of the clypeal suture obvious *Macropelopia nebulosa*
b Lateral setae on segment VII maximum 4-9 (mean 6.6). Number of setae posteriorly of the two robust setae: 1 or 2. No yellow band along the suture, therefore the suture inconspicious *Macropelopia notata*

13 a Tentorial line strongly pigmented. Tentorial pit seems to be absent (fig 28). Postoccipital margin dorsally uniformly pigmented and not broadened (as in fig 28 - ↑) *Apsectrotanypus trifascipennis*
b Tentorial line not pigmented. Tentorial pit present. Postoccipital margin dorsally broadened, not uniformly pigmented and with a dark median line (fig 29) *Macropelopia*

14 Head elongate. Abdominal segments without lateral fringe of setae.
a Anal tubules very long: $^{3}/_{4}$ the length of the posterior parapod or more (fig 32 - TA). Middle tooth of ligula longer than adjacent teeth (fig 33) **15**
b Anal tubules at most $^{1}/_{2}$ the length of the posterior parapod. Middle tooth of ligula as long as or shorter than adjacent teeth **16**

K

35 *Labrundinia* –
claw of posterior
parapod

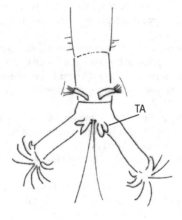

TA

37 *Guttipelopia guttipenn*
variation in eye spot

36 *Guttipelopia guttipennis* –
segment VIII-X dorsally

38 *Guttipelopia guttipennis* –
claws of posterior parapod

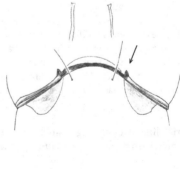

39 *Ablabesmyia
monilis* –
postoccipital
region ventrall

40 *Ablabesmyia longistyla* –
maxillary palp

41 *Ablabesmyia monilis* – maxillary palp

15 a Head margin behind the eye spots broadened and with a group of small spines (fig 34 - ↑). Abdominal setae $^1/_4$ of the segment width. Antennal segment 2 weakly pigmented. One small claw of the posterior parapod bifid (fig 35). Small species: instar iv 4.5-5.0 mm *Labrundinia longipalpis*

b Haed of normal shape and without spines. Abdominal setae about $^1/_2$ of the segment width. Antennal segment 2 not pigmented. Small claws of the posterior parapod simple. Very small species: instar iv about 3.0 mm *Nilotanypus dubius*

16 a Anal tubules short, not or barely longer than wide (fig 36 - TA). Usually two eye spots, one large upper and one small lower spot (sometimes only on one side of the head) (fig 37). One or two small claws of the posterior parapod darkly pigmented and three small not pigmented claws with a comb of teeth (fig 38, photo 14) *Guttipelopia guttipennis*

b Anal tubules more than twice as long as wide. One eye spot, usually with small incision. Some small claws of the posterior parapod can be darkened **17**

17 a Two small claws of the posterior parapod are darkened; in instar iii or ii usually weakly darkened (photo 15). All small claws smooth. Abdominal setae absent or very short, their length <1/5 of the segment width (usually shorter than in fig 36)(Examine several segments! Setae barely visible at 40x magnification). Postoccipital margin ventrally with two hooks (fig 39 - ↑, photo 16). Hooks in instar iii sometimes absent

(Ablabesmyia) **18**

b1 Postoccipital margin with two small hooks (fig 59 - ↑, photo 17) and one small claw of the posterior parapod with one tooth (fig 50) *Zavrelimyia*

b2 Small claws not pigmented, but some long claws can be weakly pigmented. If some small claws of the posterior parapod are darkened, then one or more small unpigmented claws with one long tooth or some coarse teeth. If there are small claws with one or two long teeth then go to 27. Abdominal setae length ≥ 1/4 of the segment width **21**

18 a Head length >1150 μm *Ablabesmyia phatta* instar iv

b Head length < 1150 μm **19**

19 a Basal segment of maxillary palp divided into 5-6 segments (fig 40) (Best seen in ventral or lateral view and with diascopic illumination. Note that intermediate segmentation occurs!) *Ablabesmyia longistyla*

b Basal segment of maxillary palp divided into 2 segments (Note that intermediate segmentation occurs!) (fig 41) **20**

20 We give the following ranges for separating the species. Note that there is some overlap.

a **Head length 700-1150 μm:**
Head length 900-1100(?1150) μm *Ablabesmyia monilis* instar iv
Head length 700-900 μm *Ablabesmyia phatta* instar iii

b **Head length 420-700 μm:**
Head length 540-700 μm *Ablabesmyia monilis* instar iii
Head length 420-540 μm *Ablabesmyia phatta* instar ii

K

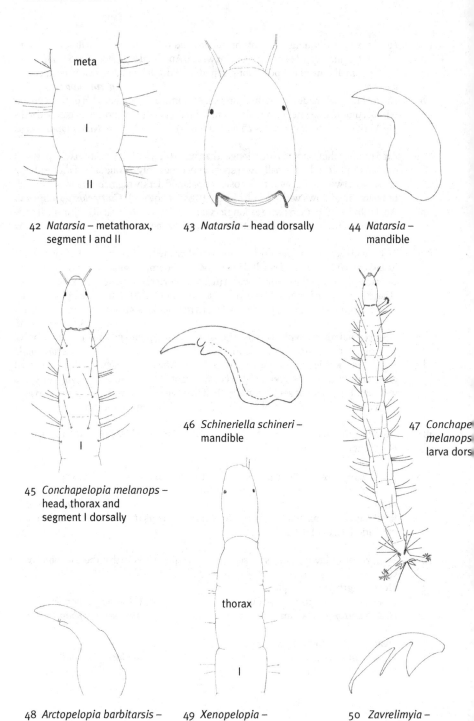

42 *Natarsia* – metathorax, segment I and II

43 *Natarsia* – head dorsally

44 *Natarsia* – mandible

45 *Conchapelopia melanops* – head, thorax and segment I dorsally

46 *Schineriella schineri* – mandible

47 *Conchape melanops* larva dors

48 *Arctopelopia barbitarsis* – mandible

49 *Xenopelopia* – head, thorax and segment I dorsally

50 *Zavrelimyia* – claw of posterior parapod

21 a Abdominal segments I-VII with a short row of 4 anterolateral setae, usually two moderately long and two very long setae (fig 42 – I , II, photo 18). Head short: IC 0.65-0.70
(fig 43). Mandible with a very large basal tooth and a very small, hardly or not visible
accessory tooth (fig 44)   ***Natarsia***

b Abdominal segments I-VII usually with one anterolateral seta. If there are 3 anterolateral setae, then never in a row (fig 45 - ↑). Head relatively more elongate: IC 0.45-0.63
(fig 45). If mandible with basal tooth large than also accessory tooth obvious (fig 46) **22**

22 a Thorax and abdominal segments I-VII dorsally no long setae. Metathorax with 2 relatively short lateral setae (fig 49). Setae of thorax and abdominal segments I-VII relatively short with length maximum $^1/_2$ of the segment width, except in *Paramerina*.
Mandible basal tooth large and accessory tooth obvious (fig 46)   **23**

b Thorax and abdominal segments I-VII dorsally with long setae. Metathorax with 3
long lateral setae; usually one seta longer than segment width and two moderately
long setae (fig 45). Setae of thorax and abdominal segments I-VII relatively long with
length $^1/_2$ - $^3/_4$ of the segment width or longer (fig 47). Mandible basal tooth and
accessory tooth very small (fig 48)   ***Conchapelopia*** agg **35**

23 a Head length 1000-1200 μm   ***Trissopelopia longimanus*** instar iv
b Head length ≤1000 μm   **24**

24 a 3-5 small claws of the posterior parapod with coarse teeth or a comb of teeth. (figs 51,
53, 54)   **25**
b All small claws of the posterior parapod smooth or 1-2 small claws with one long tooth
(fig 50)   **27**

25 a Claws of the posterior parapod: some small claws with coarse teeth (fig 51); one small
claw with one or two teeth. Ligula margin of teeth concave (fig 52)
                                                              Monopelopia tenuicalcar
b Claws of the posterior parapod: some small claws with a comb of long teeth (figs 53,
54); brown claws smooth or all claws yellow. Ligula margin of teeth straight (fig 55) **26**

51 *Monopelopia tenuicalcar* –   52 *Monopelopia tenuicalcar* –   53 *Xenopelopia* –
   claws of posterior parapod   ligula and pecten   claws of posterior parapod
                                       hypopharyngis

54 *Telmatopelopia nemorum* - claws of posterior parapod

55 *Xenopelopia* – ligula

56 *Xenopelopia* – postoccipital region ventrally

57 *Telmatopelopia nemorum* - postoccipital region ventrally

58 *Krenopelopia* - postoccipital region ventrally

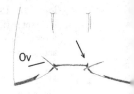

59 *Zavrelimyia* - postoccipital region ventrally

60 *Paramerina cingulata* – postoccipital margin dorsally

61 *Paramerina cingulata* – claw of posterior parapod

62 *Schineriella schineri* – postoccipital margin dorsally

63 *Schineriella schineri* – claws of posterior parapod

26 a Postoccipital margin ventrally narrow and completely pigmented (fig 56, photo 19). Antennal segment 2 (photo 20) and some small claws of the posterior parapod pigmented. (maybe inconspicuous in instar iii). Claws as in fig 53 and photo 21. Procercus length: instar iv, 170-200 μm; instar iii, 92-96 μm   ***Xenopelopia***

b Postoccipital margin ventrally broadened (only instar iv) and not pigmented, only one line obvious (fig 57, photo 22). Antennal segment 2 and all small claws of the posterior parapod yellow. Claws as in fig 54 and photo 23. Procercus length: instar iv, 120-140 μm; instar iii, 76-80 μm   ***Telmatopelopia nemorum***

27 a Head length > 550 μm
                      (instar iv larvae, also instar iii larvae of ***Zavrelimyia*** and ***Trissopelopia***) **28**

b Head length ≤550 μm   (instar iii larvae) **32**

28 a Head length 550-600 μm. Postoccipital margin pale. Ov setae of prothorax robust (fig 58, photo 24). All small claws of the posterior parapod without obvious teeth (Semi-terrestrial species)   ***Krenopelopia***
      (If head length is about 550 μm and one small claw of the posterior parapod with a long tooth   ***Zavrelimyia*** instar iii)

b Head length 600-1000 μm. Postoccipital margin pigmented. Ov setae of prothorax robust or thin   **29**

29 a All small claws of the posterior parapod without obvious tooth. Ov setae of prothorax robust (as in fig 58)   ***Trissopelopia longimanus*** instar iii and iv

b One or two small claws of the posterior parapod with a long tooth (fig 50, photo 25). Ov setae of prothorax thin, sometimes inconspicuous (fig 59 – Ov)   **30**

30 a Postoccipital margin ventrally with two small hooks (fig 59 - ↑, photo 17); in some species the margin and hooks are pale! (best seen in lateral view). Head length 800-1000 μm   ***Zavrelimyia***

b Postoccipital margin ventrally without hooks. Head length 600-780 μm   **31**

31 a Head length 600-700 μm. Abdominal setae length ¹/2 - 3/4 of the segment width. Postoccipital margin dorsally relatively narrow and completely pigmented (fig 60, photo 26). Usually antennal segment 2 brownish. One small claw of the posterior parapod brown and one small yellow claw with a long and narrow tooth (fig 61); some long claws brownish   ***Paramerina cingulata***

b Head length 730-780 μm. Abdominal setae length <¹/3 of segment width. Postoccipital margin dorsally somewhat broadened and not pigmented in the middle (fig 62). Antennal segment 2 not pigmented. No claws of the posterior parapod brownish and two small adjacent claws with a long tooth, somewhat broadened at base (fig 63, photo 27)   ***Schineriella schineri***

32 a All small claws of the posterior parapod without an obvious tooth. Head length 350-430 μm (Semi-terrestrial species)   ***Krenopelopia***

b One or two small claws of the posterior parapod with a long tooth (fig 61). Head length 420-600 μm   **33**

K

64 *Arctopelopia barbitarsis* – b
seta

65 *Conchapelopia melanops* –
b seta

66 *Rheopelopia* ?*ornata* –
pseudoradula

67 *Arctopelopia barbitarsis* –
mentum appendage with
pseudoradula

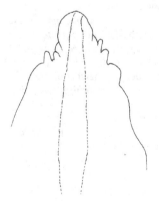

68 *Conchapelopia melanops* –
mentum appendage with
pseudoradula

69 *Thiennemannimyia* –
pseudoradula with
curled tip

33 a   Two small claws of the posterior parapod with a long tooth (fig 63). Head length 450-470 μm. Ligula teeth yellow or dark yellow   ***Schineriella schineri***

    b   One small claw of the posterior parapod with a long tooth (fig 61, photo 25). Head length 420-600 μm. Ligula teeth brownish   **34**

34 a   Head length 420-475 μm. Usually antennal segment 2 weakly pigmented. Postoccipital margin ventrally without two small hooks   ***Paramerina cingulata***

    b   Head length 480-600 μm. Antennal segment 2 not pigmented. Postoccipital margin ventrally with two extremely small hooks, almost not visible (fig 59 - ↑)   ***Zavrelimyia***

Note

The *Conchapelopia* agg includes the genera *Arctopelopia, Conchapelopia, Rheopelopia*, and *Thiennemannimyia*. The larvae of these genera are very similar and probably only distinguishable by their head length. If associated material (prepupae with thoarcic horn or pupal exuviae) is collected at the same site, identification can be more certain.

The pseudoradula can be bent or even curled at the tip (fig 66 and 69, photo 31)!

See matrix 5.4.7 for an overview of characters.

35 a   Head length 1000-1130 μm   ***Arctopelopia barbitarsis*** instar iv

    b   Head length <1000 μm. Identification is often doubtfull   **36**

36 a   Head length >650 μm   instar iv larvae and ***Arctopelopia barbitarsis*** instar iii **37**

    b   Head length 390-650 μm   instar iii larvae **40**

37 a   Four small claws of the posterior parapod brown (may fade when larvae are in alcohol for a long time). Claws of anterior parapod with relatively long inner teeth (photo 28); small claws with 1 or 2 and midlong claws with 5-9 inner teeth. Ov setae and abdominal setae very robust (photo 29), their width approximately 5 μm, in one species 4 μm. Maxillary palp b seta with basal segment divided (as in fig 65, photo 30). Pseudoradula gradually narrowing from base to tip (fig 66, photo 31). Head length 650-820 μm (Only in rivers)   ***Rheopelopia***

    b   Small claws of the posterior parapod at most brownish. Claws of anterior parapod with relatively short inner teeth (photo 32); small claws with 1-2 and midlong claws with 3-7 inner teeth. Ov setae and abdominal setae usually 3 μm, maximum 4 μm (at base sometimes more than 4 μm) (photo 33). Maxillary palp b seta with basal segment divided or not divided. Pseudoradula may or may not narrowing to the tip (figs 67, 68, 69). Head length 650-1000 μm   **38**

38 a   Usually some claws of the posterior parapod more darkened (brownish-yellow). Head length <670 μm. Maxillary palp b seta with basal segment not divided (fig 64 - ↑, photo 34). Pseudoradula in the middle more abruptly narrowing (fig 67 - ↑, photo 35)   ***Arctopelopia barbitarsis*** instar iii

    b   Claws of posterior parapod equally coloured (yellow or dark yellow). Head length 750-1000 μm. Maxillary palp with basal segment divided or not divided. Pseudoradula gradually narrowing from the middle to the tip (figs 68, 69)   **39**

39 a Maxillary palp b seta with basal segment divided (fig 65 - ↑, photo 36). Pseudoradula a broad band, weakly narrowing at the tip (fig 68). (Very common in brooks, rivers and canals)   *Conchapelopia melanops*

b Maxillary palp b seta with basal segment not divided (as in fig 64, 37). Pseudoradula a uniformly broad band (fig 69, photo 38). (Species of fast flowing lowland brooks, very rare in the Netherlands)   *Thiennemannimyia*

Note: Rarely *A. barbitarsis* with headlength 900-1000 µm.

40 a Head length 600-670 µm. Pseudoradula in the middle more abruptly narrowing (fig 67)   *Arctopelopia barbitarsis* (instar iii)

b Head length 390-590 µm   **41**

41 **The following species, all instar iii larvae, are difficult to separate. These characters can probably be used for identifying.**

a Head length 390-492 µm. Probably pseudoradula gradually narrowing from the middle to the tip (fig 66). Maxillary palp b seta with basal segment divided (Common in rivers on stones)   *Rheopelopia*

b Head length 450-520 µm. Probably pseudoradula weakly narrowing (fig 68). Maxillary palp b seta with basal segment divided (Very common in brooks, rivers and canals)

     Conchapelopia melanops

c Head length 474-588 µm. Probably pseudoradula uniformly broad (fig 69). Maxillary palp b seta with basal segment not divided (Species of fast flowing lowland brooks, very rare)   *Thiennemannimyia*

5.2 KEY TO PREPUPAE

Thoracic horns can be used to identify Tanypodinae.

At the end of the fourth larval instar the larva develops into a pupa. The relatively short period in which the larva changes into a pupa is called the prepupa.

The prepupa can easily be recognized because the thorax is obviously thickened and some parts of the adult body develop just beneath the larval skin. For respiration as a pupa the larva develops two thoracic horns, which can be observed at the end of the prepupa's development into a pupa. The thoracic horns are partly or almost completely pigmented and can easily be observed because the skin is transparent (photo 48 and 49). If the details of the thoracic horn are difficult to identify, it is advisable to dissect one thoracic horn by opening the skin carefully. The thoracic horn can also be mounted on a slide for examination under the microscope.

The species can be identified by examining the pupa or pupal exuviae. In keys to the pupae the thoracic horn is an important feature, but not the only one! In many cases, though, it is possible to identify species of the Tanypodinae by examining the thoracic horn.

In this "key" the thoracic horns are placed in groups. In some genera, however, the shape of the thoracic horn differs very little between species. In such case it is only possible to identify the genus. In the key these genera are indicated as follows: *Xenopelopia* (figure: Langton: 1991 – *X. falcigera*)

GROUPS OF THORACIC HORNS

Thoracic horn: bulbous and brown pigmented

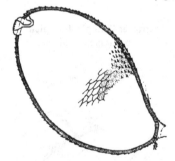

Ablabesmyia
(figure: Langton: 1991 – *A. longistyla*)
ThL: 370-625 µm: *longistyla/monilis*
ThL: 815-880 µm: *phatta*

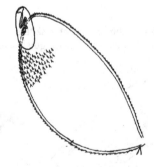

Guttipelopia guttipennis
(figure: Langton: 1991)
ThL: 585-705µm

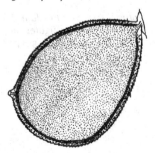

Procladius crassinervis
(figure: Langton: 1991)
ThL: 730-1030µm

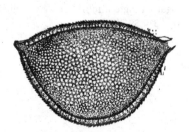

Tanypus punctipennis
(figure: Langton: 1991)
ThL: 645-885µm

43

Thoracic horn: form of a skittle. Respiratory atrium wider than plastron plate

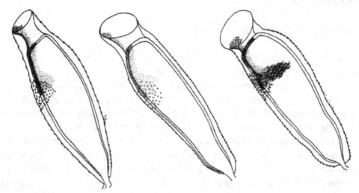

Procladius choreus
(figure: Langton: 1991)
ThL: 385-590 µm

Procladius Pe4
(figure: Langton: 1991)
ThL: 370-495 µm

Procladius signatus/sagittalis
(figure: Langton: 1991 –
P. signatus)
ThL: unknown

Thoracic horn: form of a skittle. Respiratory atrium narrower than plastron plate
It looks possible to distinguish these species by the form, but it is not reliable.

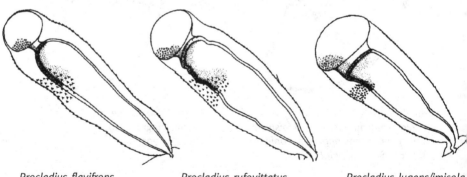

Procladius flavifrons
(figure: Langton: 1991)
ThL: 310-395 µm

Procladius rufovittatus
(figure: Langton: 1991)
ThL: 385-490µm

Procladius lugens/imicola
(figure: Langton: 1991)
ThL: 565, 580µm

Thoracic horn: divided into two parts by respiratory atrium and plastron plate

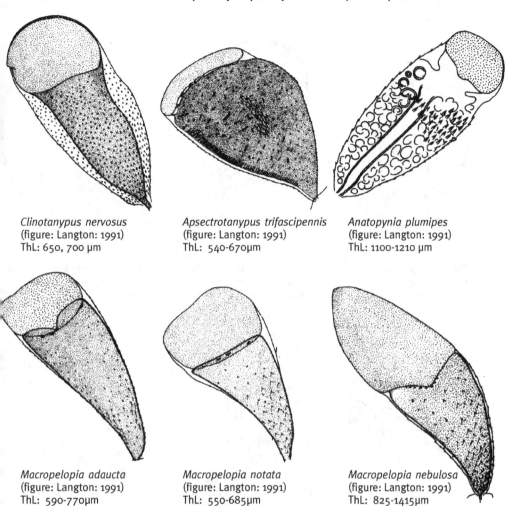

Clinotanypus nervosus
(figure: Langton: 1991)
ThL: 650, 700 μm

Apsectrotanypus trifascipennis
(figure: Langton: 1991)
ThL: 540-670μm

Anatopynia plumipes
(figure: Langton: 1991)
ThL: 1100-1210 μm

Macropelopia adaucta
(figure: Langton: 1991)
ThL: 590-770μm

Macropelopia notata
(figure: Langton: 1991)
ThL: 550-685μm

Macropelopia nebulosa
(figure: Langton: 1991)
ThL: 825-1415μm

Note: The species *M. adaucta* and *M. notata* cannot be distinguished by the shape of the thoracic horn!

Thoracic horn: club shaped

without rim   with rim   with rim

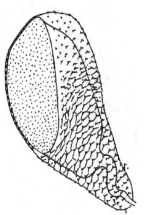

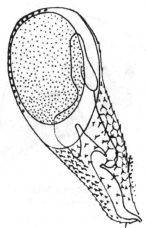
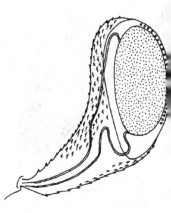

Conchapeolpia melanops
(figure: Langton: 1991)
ThL: 385-560 µm

Conchapelopia pallidula
(figure: Langton: 1991)
ThL: 310-410 µm

Conchapelopia hittmairorum
(figure: Langton: 1991: Pe1)
ThL: 285-395µm

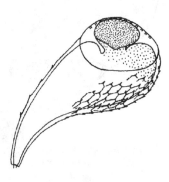
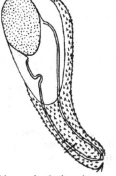

Telopelopia fascigera
(figure: Langton: 1991)
ThL: 350, 360µm

Trissopelopia longimanus
(figure: Langton: 1991)
ThL: 345-420 µm

Labrundiria longipalpis
(figure: Langton: 1991)
ThL: 295 µm

Thoracic horn: a cylindrical tube. Plastron plate without rim

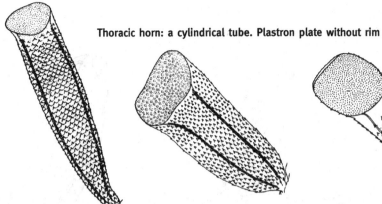

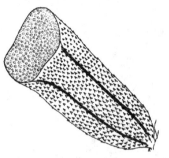

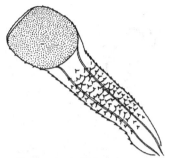

Psectrotanypus varius
(figure: Langton: 1991)
ThL: 565-770µm

Natarsia (figure: Langton: 1991
– *N. nugax*)
ThL: 380, 420 µm

Krenopelopia (figure: Langton:
1991 – *K. nigropunctata*)
ThL: 320-440µm

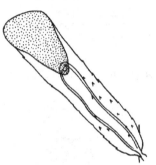

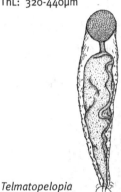

Monopelopia tenuicalcar
(figure: Langton: 1991)
ThL: 265-395 µm

Telmatopelopia nemorum
(figure: Langton: 1991)
ThL: 430µm

*Telmatopelopia
nemorum*
(figure: after Fittkau: 1960)

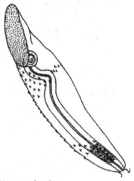

Schineriella schineri
(figure: Langton: 1991)
ThL: 375µm

Zavrelimyia nubila
(figure: after Fittkau: 1960)
ThL: 350-450µm

Xenopelopia
(figure: Langton: 1991 –
X. falcigera)
ThL: 505, 515µm

Notes:
In *Schineriella schineri* the atrium nearly filling the horn lumen and
the thoracic horn without reticulation. In *Zavrelimyia* the atrium is
less wide and the thoracic horn with reticulation.
The characters given by Pankratova (1977) and Langton (1991) are not
correct, so it is not possible to separate the species.

Thoracic horn: a cylindrical tube. Plastron plate with rim

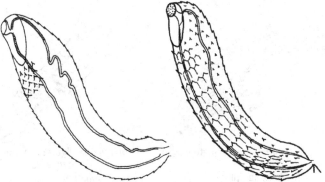
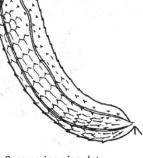
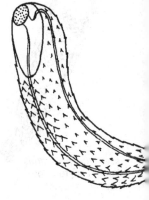

Zavrelimyia hirtimana
(figure: Langton: 1991)
ThL: 400-450μm

Paramerina cingulata
(figure: Langton: 1991)
ThL: 225-330 μm

Paramerina divisa
(figure: Langton: 1991)
ThL: unknown

Thoracic horn of *Zavrelimyia hirtimana* and *Paramerina cingulata* are very similar!

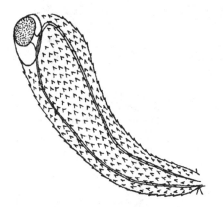

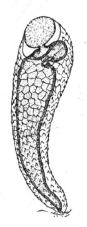

Zavrelimyia melanura
(figure: Langton: 1991)
ThL: 345-425μm

Larsia
(figure: Langton: 1991 –
L. atrocincta)
ThL: 265μm

Zavrelimyia signatipennis
(after Fittkau: 1962)
ThL: 316-383 μm

Nilotanypus dubius
(figure: Langton: 1991)
ThL: 170-215μm

Thoracic horn: shape simple

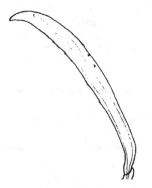

Thienemannimyia carnea
(figure: Langton: 1991)
ThL: 230-290 µm

Thienemannimyia lentiginosa
(figure: Langton: 1991)
ThL: 300-385 µm

*Thienemannimyia
pseudocarnea*
(figure: Langton: 1991)
ThL: 220-345µm

Rheopelopia
(figure: Langton: 1991 –
R. ornata)
ThL: 310, 345µm

Arctopelopia
(figure: Langton: 1991 – *A. griseipennis*)
ThL: 345-405 µm

5.3 COMMENTS

This book does not contain complete descriptions of the larvae. The reader can use these comments, which refer only to the most distinguishing characters, to check identification. Brief descriptions of the larvae can be found in Moller Pillot (1984); for more complete information and many figures, see Fittkau and Roback (1983). For an overview
of some characters see matrices 1-7.
Wide headed larvae: matrix 1.
Elongate headed larvae, measurements: matrix 2 and characters: matrix 3.
Shape of head in wide headed larvae: matrix 4; in elongate headed larvae: matrix 5.
Postoccipital region ventrally of elongate headed larvae: matrix 6.
Conchapelopia aggregate: matrix 7.

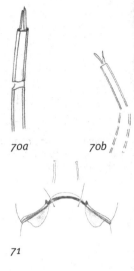

70a   70b

ABLABESMYIA
The genus *Ablabesmyia* is characterised by very short abdominal setae in combination with two dark claws on the posterior parapod and the division of the basal segment of the maxillary palp (fig 70a and b). The small hooks ventrally on the postoccipital margin (fig 71) are only present in *Ablabesmyia* and *Zavrelimyia*. With some experience it is possible to recognise the head shape, especially in instar iv. The larva resembles *Guttipelopia*, which has much shorter anal tubules.

71

For information about the taxonomic problems in this genus and mistakes in the literature, see Chapter 7, p62. The subdivision of the maxillary palp is variable, at least within *A. monilis*, but the basal segment may never be subdivided as in *A. longistyla*.

ANATOPYNIA PLUMIPES
This is the largest species of the Tanypodinae and superficially resembles *Psectrotanypus varius* because of its pale, rounded-oval head capsule. The species can be distinguished from all other species by the procercus, which is 2.5–3 times as long as it is wide. In instar iii and iv the procercus has more than 25 setae (in other species the procercus is longer and it has fewer than 20 setae) (photo 39). The mandible is without a basal tooth and has two accessory teeth.

APSECTROTANYPUS TRIFASCIPENNIS
This species is distinguishable from all other species by the brown head capsule with a pale spot around the eyes in instar iv larvae. Instar iii larvae, however, look similar to *Macropelopia*. *A. trifascipennis* is distinguished by the narrow pale band along the coronal suture. Besides the characters mentioned in the key, the head length can be used for identification (see matrix tables). The spotted thorax of living instar iv larvae resembles that of some *Procladius* larvae. Instar iv larvae differ from all related genera by the 4–5 unusually long dorsomental teeth (other genera have 6 or more teeth).

ARCTOPELOPIA
Among the *Conchapelopia* agg. (see 23 and 36 in the key), the larvae of *Arctopelopia* in instar iv are recognisable by their head length, which is more than 1000 μm (found within the Pentaneurini only in *Ablabesmyia* and *Trissopelopia*). A further character is the pecten hypopharyngis with 25 teeth, becoming gradually smaller laterally (in other genera of the aggregate approx. 20 teeth; see Fittkau and Roback, 1983). The larva also differs from *Rheopelopia* and *Conchapelopia* by the 2 segmented seta b of the maxillary palp (basal segment not divided).

CLINOTANYPUS NERVOSUS

The larva is distinguishable from all other genera by its wedge-shaped head when viewed laterally. Also a unique character is the presence of a small conical papilla, as tall as it is wide, located between the procerci (fig 72). Eye ventrally partly elongated. The species differs from all other wide-headed species by the dorsomental teeth, which consist of a row of approximately 15 small sharp points (photo 40).

72

Conchapelopia agg.

This aggregate is probably not a systematic unit, but the larvae are just about distinguishable as a group. The aggregate is defined by the characters mentioned in the key under 36a and in our region includes the genera *Arctopelopia*, *Conchapelopia*, *Rheopelopia* and *Thienemannimyia*.

CONCHAPELOPIA

Our key only includes the species *C. melanops* because other species live only in faster running streams and have never been found in the Netherlands. The genus is barely distinguishable from *Rheopelopia*. *C. melanops* usually has a larger head. In fresh specimens the claws of the posterior parapod are never dark brown. The pseudoradula barely narrows from base to tip. The larva can be separated from *Arctopelopia* and *Thienemannimyia* by the 3 segmented b seta of the maxillary palp (basal segment divided).

73

GUTTIPELOPIA GUTTIPENNIS

The small and short anal tubules, no longer or barely longer than they are wide, are unique within the Tanypodinae. Another unique character is the presence of fine longitudinal striae on the thorax and abdomen. Head capsule with fine granulation. Living larvae are greenish, with many small white spots, but lose their colour in alcohol.

74

KRENOPELOPIA

Krenopelopia is easily recognisable by the large teeth of the mandible and the relatively wide head capsule (IC > 0.60). Because of their small size, the instar iv larvae only resemble instar iii larvae of related genera. In contrast to *Natarsia*, the thorax and abdomen of the living larva are white.

LABRUNDINIA LONGIPALPIS

This species is unique among the European Tanypodinae in having a hump with small points behind the eyes (fig 73). Further characteristic features are the bifid claw on the posterior parapods (fig 74) and the seta which arises ventrally from the basal part of the posterior parapods bearing small spines (fig 75).

75

MACROPELOPIA

Instar iv larvae differ from all related (wide-headed) genera by the slightly longer, orange head capsule (slightly brownish in instar iii) and the blood-red colour of the full-grown living larvae. *Procladius* specimens with a slightly orange head capsule can resemble *Macropelopia*, but have a black ligula. Instar iii larvae resemble those of *Apsectrotanypus*. *Macropelopia* is distinguished from *Procladius* and P*sectrotanypus* by the narrow pale band along the coronal suture. The pale band in M. nebulosa obvious, in M. adaucta and M. notata inconspiciuous.

MONOPELOPIA TENUICALCAR

The larva is distinguishable from other small species with a dark second antennal segment and some dark claws on the posterior parapod (*Xenopelopia, Paramerina cingulata*) by the typical teeth on some small claws of the posterior parapod. These claws are unique within the Tanypodinae.

NATARSIA

The genus is easily identifiable by the very long anterolateral setae and the very large basal tooth of the mandible. *Telopelopia fascigera*, a rare species inhabiting large European rivers, also has a large basal tooth on the mandible. This species very probably also has long dorsal abdominal setae, like the other members of the *Conchapelopia* aggregate. In common with *Natarsia, Krenopelopia* has a large basal tooth, but also has a well developed accessory tooth. Living instar iv larvae can be identified by the relatively short head (IC nearly 0.7) in combination with the red body colour.

NILOTANYPUS DUBIUS

Supra-anal setae and anal tubules as long as the posterior parapods (fig 76). Some small claws of posterior parapods with a comb of teeth. Only in fast flowing streams.

76

PARAMERINA CINGULATA

A small larva with a strikingly narrow head (IC ± 0.5). Thorax of living larvae in instar iii and iv with a pattern of dark reddish-brown lines and spots. The larva has a dark second antennal segment and two dark claws of the posterior parapod in common with *Monopelopia* and *Xenopelopia*, which differ most noticeably in the teeth of the small pale claws. Only one claw of the posterior parapods has a long and narrow tooth as in *Zavrelimyia*.

PARAMERINA DIVISA

This species is unlikely to be present in the Netherlands and adjacent lowlands. The larva differs noticeably from *P. cingulata* because all claws of the posterior parapod are yellow and simple. For a description see Laville (1971). Moller Pillot (1984) writes incorrectly that the second antennal segment is pale.

PROCLADIUS

Procladius differs distinctly from other wide-headed genera by the black ligula and the absence of a lateral fringe of long setae on abdominal segment VII. The subgenera cannot yet be identified in reliable manner. The characters mentioned by Fittkau & Roback (1983) probably do not apply to all species.

The colour of the thorax and abdomen of living larvae is very variable, also within one species. The thorax is often spotted; haemoglobin may be present, but the larva never looks blood red. At first sight *Procladius* is often recognisable by a (paired or unpaired) dark spot dorsally on the back side of the head (see photo on the cover).

PROCLADIUS (PSILOTANYPUS) RUFOVITTATUS

A small claw of the posterior parapods has obvious inner teeth (fig 77).

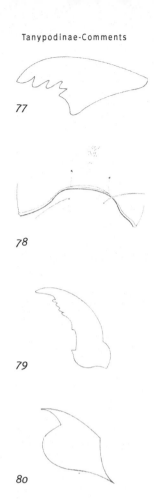

77

PSECTROTANYPUS VARIUS

Head colour pale yellow. Submentum with a group of spinules (fig 78, photo 41). In contrast to all other wide-headed larvae, the species is characterised by the ligula with 4 equally long and yellow teeth and the mandible in instar iii with a row of 2-4 inner teeth, in instar iv with 4-6 inner teeth (fig 79). One small claw of the posterior parapod very typical (fig 80).

78

RHEOPELOPIA

The larva resembles that of *Conchapelopia*. For comments see under this genus. In some cases, head length can be decisive: in *Rheopelopia* 650–820 μm, in *Conchapelopia melanops* 750–920 μm.

SCHINERIELLA SCHINERI

The larva of *Schineriella schineri* resembles those of *Zavrelimyia* and *Paramerina cingulata*. Common characters among these genera are the narrow head, the long tooth on one or two claws of the posterior parapod and the equally long teeth of the ligula. In *Schineriella* two claws always have this long tooth, which is broader at the base than in both other genera. Besides the characters mentioned in the key, the yellow to brownish-yellow teeth of the ligula (never dark brown as in both other genera) are characteristic for *Schineriella* (photo 42). In instar iii and iv the head length is also a useful character.

79

80

K

TANYPUS

This genus has 4 or 6 anal tubules. The larvae look similar to those of *Procladius* and *Psectrotanypus*, especially in *T. vilipennis*, where the number of the anal tubules is equal to those of all other Tanypodinae. Tentorial pit with a dark ingrowth; tentorial line thin and relatively very short (fig 81). Segment VI (and sometimes also segment VII) in instar iv has a bulbous lateral extension (fig 82). Two small claws of posterior parapod with bulbous base (fig 83). Ligula with 5 equally long yellow teeth (photo 43). The shape of the mandible is characteristic: an enlarged base and small basal and very small accessory teeth (fig 84, photo 44).

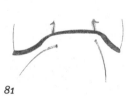

81

TANYPUS KRAATZI

Head pale. Dorsomentum with usually 6 (rarely up to 8) brown lateral teeth (fig 85 and photo 45); in instar iii 4–6 brownish teeth. Paraligula with two long branches (fig 86).

TANYPUS PUNCTIPENNIS

Head pale. Dorsomentum with 7–9 short pale teeth; in instar iii 6–7 (fig 87). Paraligula with about 13 branches (fig 88, photo 12).

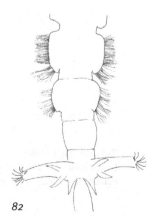

82

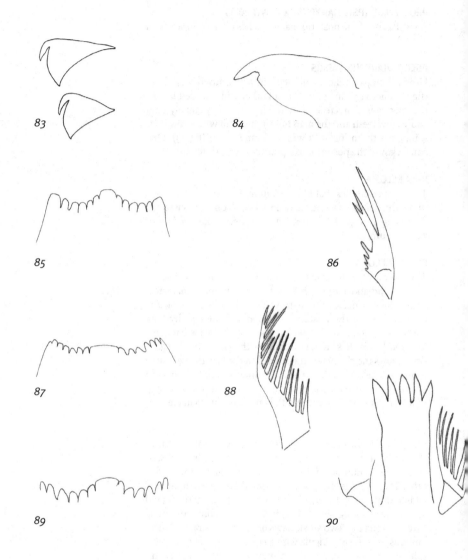

TANYPUS VILIPENNIS

In contrast to other *Tanypus* species, only 4 anal tubules. 5–8 dorsomental teeth (fig 89). Paraligula with about 6 branches (fig 90, photo 46).

TELMATOPELOPIA NEMORUM
The larva closely resembles *Xenopelopia*, but the living larva can be recognised by the pink coloured abdomen. It differs from *Xenopelopia* and *Monopelopia* by the head size and because antennal segment 2 and claws of the posterior parapod are never darkened. All other species, except for *Guttipelopia* and *Nilotanypus*, lack the comb of long teeth on some small claws of the posterior parapod.

TELOPELOPIA FASCIGERA
A rare species of large rivers, related to *Conchapelopia*, but easily recognisable by the large basal tooth of the mandible. Not included in the key. See Fittkau and Roback (1983).

THIENEMANNIMYIA
The larva closely resembles *Arctopelopia*, but can be distinguished in instar iv by their smaller size and the lower number of teeth in the pecten hypopharyngis. However, most species of *Thienemannimyia* are not yet known as a larva and possibly there will be an overlap with *Arctopelopia*. The larva differs from *Conchapelopia* and *Rheopelopia* by the 2 segmented seta b of the maxillary palp (basal segment not divided). The larva can be separated from other Pentaneurini by the body setae and mandible, as mentioned in the key under 22.

TRISSOPELOPIA LONGIMANUS
In instar iv *T. longimanus* nearly always has a head length of more than 1.0 mm, which is found among the Pentaneurini only in *Ablabesmyia* and *Arctopelopia*. The larva of *Trissopelopia* can be separated from both these species by the conspicuous but not very long body setae, as mentioned in the key. Moreover, in *Ablabesmyia* the basal segment of the maxillary palp is divided and two claws of the posterior parapod are dark brown. Further, the instar iv larva differs from that of *Arctopelopia*, *Krenopelopia* and *Natarsia* by the slender maxillary palp (basal segment 5–6 times as long as wide). In instar iii the claws have to be studied. *Trissopelopia* lives only in springs and spring brooks. See Chapter 7 (p. 119) for information about the related *T. flavida*.

XENOPELOPIA
The narrow head (IC 0.45–0.50) and the spotted thorax are striking in the living larva, but *Xenopelopia* has these characters in common with such genera as *Zavrelimyia* and *Telmatopelopia*. The claws of the posterior parapod are characteristic for both *Xenopelopia* and *Telmatopelopia* (more or less different combs of teeth are present in *Nilotanypus*, *Guttipelopia* and *Monopelopia*). *Xenopelopia* and *Telmatopelopia* can be distinguished by the characters mentioned in the key under 27.

ZAVRELIMYIA
Zavrelimyia is easily distinguishable from the two other genera with a long tooth on a claw of the posterior parapods, at least in instar iv, by the small hooks on the ventral part of the post-occipital margin. The teeth of the ligula are dark brown in *Zavrelimyia* (yellow to brownish-yellow in *Schineriella*, photo 47). Head length is also a useful character (see matrix table). In contrast to *Schineriella*, only one claw of the posterior parapods has a long tooth, and this tooth is not broadened at the base. In samples with many *Conchapelopia* or *Ablabesmyia* larvae, the differences in head shape are a useful distinguishing feature at first glance.

5.4 MATRICES

5.4.1 WIDE HEADED LARVAE

key nr		instar	colour living larvae	hemogl living larvae	thorax spotted	lat setae thorax pro n	lat setae thorax meso n
02a	Clinotanypus nervosus	iv	red	strong	absent	1	>60
02a	Clinotanypus nervosus	iii	reddish	present	absent		30-60
04	Tanypus	iv					
04	Tanypus	iii					
04	Tanypus	ii					
04a	Tanypus kraatzi	iv	greenish	present	absent	1	3-4
04a	Tanypus kraatzi	iii		present	absent		
04b	Tanypus punctipennis	iv				16	10
04b	Tanypus punctipennis	iii					
05a	Procladius - all subgenera	iv	variable	present	present/absent	1	4-10
05a	Procladius - all subgenera	iii					
05a	Procladius - all subgenera	ii					
06a	Anatopynia plumipes	iv	brownish		absent	1	8-10
06a	Anatopynia plumipes	iii				1	3-7
07a	Psectrotanypus varius	iv	greenish	absent	absent	6-13	6-12
07a	Psectrotanypus varius	iii	rose	present	absent	3-5	8-12
07a	Psectrotanypus varius	ii					
08a	Tanypus vilipennis	iv				1	3
10a	Apsectrotanypus trifascipennis	iv	brownish		present	1	8-10
11	Macropelopia	iv	red	strong	absent		
11a	Macropelopia adaucta	iv				1	15-21 (2) (
12a	Macropelopia nebulosa	iv				1	8-10 (2) 3
12b	Macropelopia notata	iv				1	5-9 (2) o
13	Macropelopia	iii	pale	absent?	absent		
13	Macropelopia	ii					
13a	Apsectrotanypus trifascipennis	iii					
13b	Macropelopia nebulosa	iii				4-7	
13b	Macropelopia nebulosa	ii					
13b	Macropelopia notata	ii					

The ranges given in matrix 1, 2 and 3 are from our own measurements. They are probably not exact because of the low numbers of larvae examined. We have many measurements of head length and only a few of head width. Considerable deviations in size may occasionally be encountered.

Note that setae can be broken off.

Abbreviations:

inc = inconspicious

[ ] sizes based on the rule of 60%

hemogl = hemoglobine. This may give the abdomen a colour of red.

lat setae thorax meta n	lat.fringe VII setae n	head length µm	head width µm	TeP inter distance µm	procercus length µm	anal setae n
>60	>50	950-1200	750-880	230-260	250-280	
30-50	18-20	550-700		130		
		550-650				
		370-420				
		[220] 250				
6-7	25-36	510-650	450-530	188-204	360-440	
		360-380	300			
20	32-36	560-570	510-550	200-240	320-350	
		360	320			
6-24	1-3 short	?550-820(1000)	750-830		190-280	12-18
	1-3 short	?370-490[600]				
	1-3 short	[220]290				
26-30	>60	1200-1400	1140-1180	350-390	460-500	±25
9-18	>60	750-860	700-730	220-250	250-270	25-28
12-15	16-19	740-850(900)	620-690	220	360-390	18-20
9-15	8-13	470-550	400-440		220-250	
		[280]320-340				
6	17-28	570-600	450-500	190-230	330-360	
12-20	17-22	750-880	510-680	160-180	240-260	11-14
		(850)950-1260				12-13
16-32	3-13	(1000)1140-1240	750-950	240-280	320-380	
25-36	11-32	970-1120	740-900	250	250-320	
17-18(23)	4-9	900-1160	720-770	200-230	250-280	
		560-710				
		330-350[400]	330-350[400]			
		410-500	350-400	110-115	110-130	
		590	450		180	
		350-390				
		[280]320-340				

5-4-2   ELONGATE HEADED LARVAE-MEASUREMENTS

key nr		instar	head length µm	head width µm	procercus length µm	
15	Labrundinia longipalpis	iv	500		130-140	
16	Guttipelopia guttipennis	iii	430-450			
16	Guttipelopia guttipennis	iv	650-760		140	
19	Ablabesmyia longistyla	iii	540-670			
19	Ablabesmyia longistyla	iv	900-1100		120-130	
20	Ablabesmyia monilis	iii	540-700			
20	Ablabesmyia monilis	iv	900-1100(1150)		160-170	
20	Ablabesmyia phatta	iii	700-900			
20	Ablabesmyia phatta	iv	1160-1500		160-190	
21	Natarsia	iii	450-500			
21	Natarsia	iv	700-800		240-260	
23,29	Trissopelopia longimanus	iv	1100-1200	580-625	160	
25	Monopelopia tenuicalcar	iii	430-450		70	
25	Monopelopia tenuicalcar	iv	(450)580-670		100-130	
26	Telmatopelopia nemorum	iii	360-440		76-80	
26	Telmatopelopia nemorum	iv	(530)570-650		120-160	
26	Xenopelopia	iii	430-530		92-96	
26	Xenopelopia	iv	670-860		170-200	
28	Krenopelopia	iv	550-600		80-90	
29	Trissopelopia longimanus	iii	600-650[700]		90	
30	Zavrelimyia	iv	800-1000			
30	Zavrelimyia melanura	iv	800-820		160-180	
30	Zavrelimyia nubila	iv	(780)900-920		200	
31	Paramerina cingulata	iv	600-700		130-150	
31	Schineriella schineri	iv	700-780	330-360	150-200	
32	Krenopelopia	iii	350-430		45-50	
33	Schineriella schineri	iii	[450]470			
34	Paramerina cingulata	iii	420-450			
34	Zavrelimyia	iii	530-600			
35	Arctopelopia barbitarsis	iv	1000-1130		120-150	
37	Rheopelopia	iv	650-820		100	
38	Arctopelopia barbitarsis	iii	[600-670]			
39	Conchapelopia hittmairorum	iv	740-810		210-310	
39	Conchapelopia melanops	iv	750-920		120-130	
39	Thienemannimyia	iv	980			
41	Conchapelopia melanops	iii	450-520			

ipra anal ta ngth μm	antenna length segm 1 μm	antenna length segm 2-5 μm
300	190-200	70
450-500	270	40
330-380	400	90
400	330	80
580-590	610-700	80-110
380	120-160	60-72
370	380	110
200-210	160-170	60-70
220-240	240-250	70
120	108-116	52-54
160-170	176-200	60
220	180	60-76
260-270	270-300	80-90
210-230	140-160	60
300	190	60
300-360	280-320	90-100
240-280	330-340	90
250	210-230	80
260-300	270-320	90-100
150-180	80-100	40
680-800	310-360	60-90
650	150	?
	210-310	
620-720	300	60

5·4·3   ELONGATE HEADED LARVAE-CHARACTERS

key nr		instar	colour living larvae	hemogl living larvae	thorax spotted	anal tub post par index	abdome setae length ratio seg
15	Labrundinia longipalpis	iv				$3/4$	<1/
15	Nilotanypus dubius	iv				$3/4$	$1/2$
16	Guttipelopia guttipennis	iv	greenish	absent	present	$<1/4$	<1/
16	Guttipelopia guttipennis	iii		absent		$<1/4$	<1/
18,20	Ablabesmyia phatta	iv		absent		$1/4-<1/2$	<1/
18,20	Ablabesmyia phatta	iii		absent		$1/4-<1/2$	<1/
19	Ablabesmyia longistyla	iv	yellowish	absent	present	$1/4-<1/2$	<1/
19	Ablabesmyia longistyla	iii		absent		$1/4-<1/2$	<1/
20	Ablabesmyia monilis	iv		absent		$1/4-<1/2$	<1/
20	Ablabesmyia monilis	iii		absent		$1/4-<1/2$	<1/
21	Natarsia	iv	red	strong	absent	$1/4-<1/2$	$1/2$ (4 Ls
21	Natarsia	iii	rose	present	absent	$1/4-<1/2$	$1/2$ (4 Ls
23	Trissopelopia longimanus	iv	brownish	absent	absent	$1/4-<1/2$	$1/2$
25	Monopelopia tenuicalcar	iv	yellowish	absent	absent	$1/4-<1/2$	$1/2$
26	Telmatopelopia nemorum	iv	rose	present	present	$1/4-<1/2$	$<1/2$ (+1
26	Xenopelopia	iv	brownish	present?	present	$1/4-<1/2$	$1/2$
26	Xenopelopia	iii				$1/4-<1/2$	$1/2$
28	Krenopelopia	iv	white	absent	absent	$1/4-<1/2$	$1/3$
30	Zavrelimyia	iv	brownish	absent?	present	$1/4-<1/2$	?$1/3$
30	Zavrelimyia melanura	iv				$1/4-<1/2$	?$1/3$
30	Zavrelimyia nubila	iv				$1/4-<1/2$	?$1/3$
31	Paramerina cingulata	iv	brown	absent	present	$1/4-<1/2$	$1/2-3/$
31	Schineriella schineri	iv				$1/4-<1/2$	<1/
35	Arctopelopia barbitarsis	iv				$1/4-<1/2$	$1/2-3/4$+th
37	Rheopelopia	iv				$1/4-<1/2$	$3/4$+the
39	Conchapelopia melanops	iv	white	present	absent	$1/4-<1/2$	$3/4$+the
39	Thienemannimyia	iv				$1/4-<1/2$	$1/3-1/2$+t
41	Conchapelopia melanops	iii	white	absent	absent	$1/4-<1/2$	$3/4$+the

- rax v tae	small claw post-par pigm **n**	small claw post-par tooth **n**	small claw post-par comb/coarse **n**	antenna segment2 **pigment**	ligula teeth **margin**	mandible basal tooth **size**	mandible acces tooth **size**
	0	1	0	**brownish**	**>Mi**	++	+
		longer	1		**>Mi**	++	++
a-inc	1-2	0	3	no	concave	±	+
a-inc					concave	±	+
in	2	0	0	no	concave		
in					concave		
in	2	0	0	no	concave	+	±
in					concave		
in	2	0	0	no	concave		
in					concave		
in	0	0	0	no	concave	++	--
in					concave	++	--
ust	0	0	0	no	concave	+	±
in	1-3	0	**5 coarse**	**brown**	concave	+	±
in	0	0	**3(+2) comb**	no	straight	+	±
in	2	0	5 comb	brown	straight	+	±
in	2		5 comb	brown	straight	+	±
ust	0	0	0	no	**concave**	++	±
in	0	1	0	no	straight	+	±
in	0	1	0	no	straight	+	±
in	0	1	0	no	straight	+	±
in	0-2	1	0	brownish	straight	+	±
/inc	0	2	0	no	straight	+	±
ust	5	0	0	no	concave	--	--
ust	4	0	0	no	concave	-	--
ust	0	0	0	no	concave	-	-
ust	0	0	0	no	concave	-	-
ust					concave	-	

MATRIX 5.4.4   **WIDE HEADED LARVAE**
                     POSTOCCIPITAL REGION

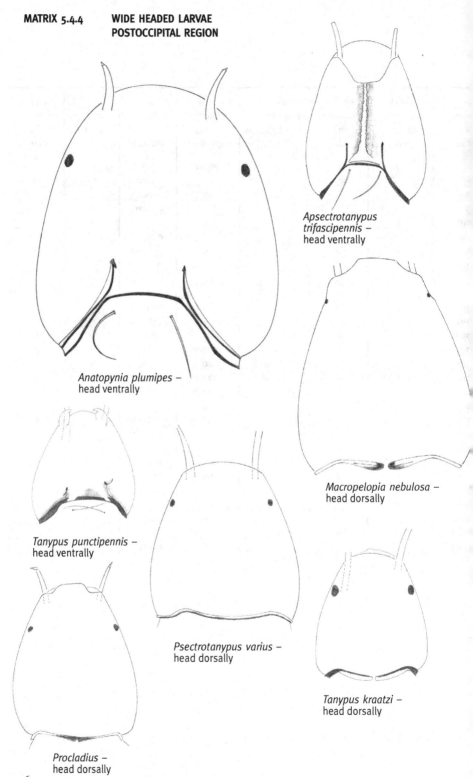

Apsectrotanypus trifascipennis –
head ventrally

Anatopynia plumipes –
head ventrally

Macropelopia nebulosa –
head dorsally

Tanypus punctipennis –
head ventrally

Psectrotanypus varius –
head dorsally

Tanypus kraatzi –
head dorsally

Procladius –
head dorsally

MATRIX 5.4.4 **WIDE HEADED LARVAE**
POSTOCCIPITAL REGION VENTRALLY

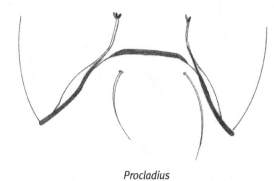

Procladius

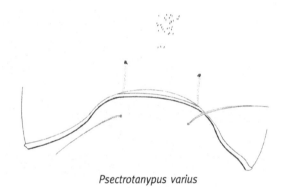

Psectrotanypus varius

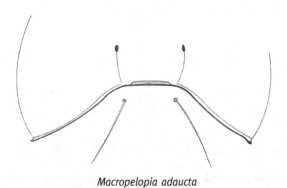

Macropelopia adaucta

MATRIX 5.4.5   **ELONGATE HEADED LARVAE
HEAD DORSALLY**

Krenopelopia

Natarsia

Trissopelopia longimanus

*Telmatopelopia
nemorum*

Thiennemannimyia

*Arctopelopia
barbitarsis*

Conchapelopia melanops
also *Rheopelopia*

MATRIX 5.4.5   ELONGATE HEADED LARVAE

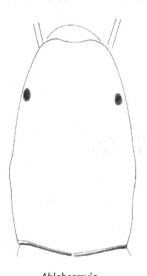

*Ablabesmyia
monilis*

*Guttipelopia
guttipennis*

Paramerina cingula-
ta

*Schineriella
schineri*

*Monopelopia
tenuicalcar*

Zavrelimyia

Xenopelopia

*Labrundinia
longipalpis*

MATRIX 5.4.6 **ELONGATE HEADED LARVAE POSTOCCIPITAL REGION VENTRALLY**

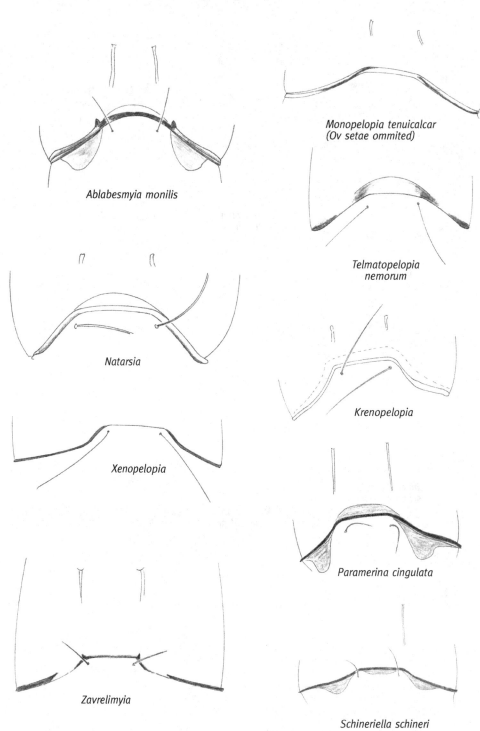

Ablabesmyia monilis

Monopelopia tenuicalcar (Ov setae ommited)

Telmatopelopia nemorum

Natarsia

Krenopelopia

Xenopelopia

Paramerina cingulata

Zavrelimyia

Schineriella schineri

MATRIX 5.4.6 **ELONGATE HEADED LARVAE
POSTOCCIPITAL MARGIN VENTRALLY
CONCHAPELOPIA AGGREGATE**

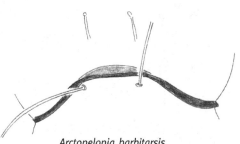

Arctopelopia barbitarsis
(Ov setae must be robust)

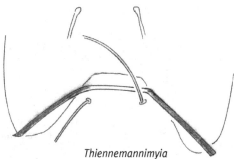

Thiennemannimyia

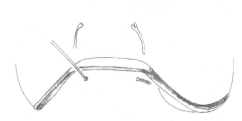

Thiennemannimyia carnea

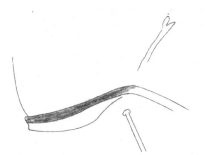

Rheopelopia maculipennis

Conchapelopia pallidula

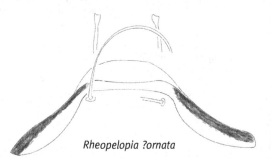

Rheopelopia ?ornata

Conchapelopia melanops

MATRIX 5.4.7 **CONCHAPELOPIA AGGREGATE**

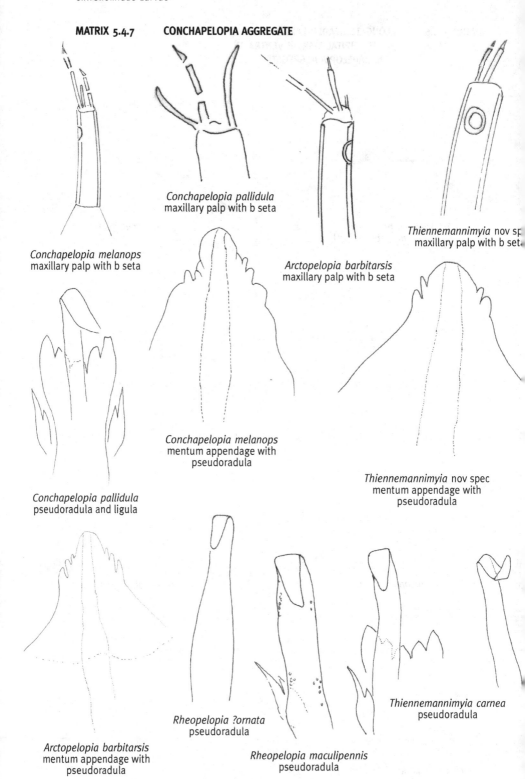

Conchapelopia pallidula
maxillary palp with b seta

Conchapelopia melanops
maxillary palp with b seta

Arctopelopia barbitarsis
maxillary palp with b seta

Thiennemannimyia nov sp
maxillary palp with b set

Conchapelopia melanops
mentum appendage with
pseudoradula

Thiennemannimyia nov spec
mentum appendage with
pseudoradula

Conchapelopia pallidula
pseudoradula and ligula

Arctopelopia barbitarsis
mentum appendage with
pseudoradula

Rheopelopia ?ornata
pseudoradula

Rheopelopia maculipennis
pseudoradula

Thiennemannimyia carnea
pseudoradula

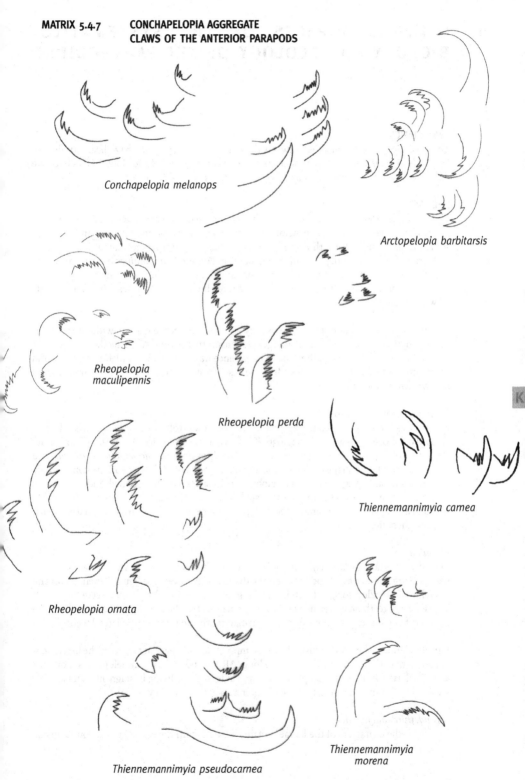

MATRIX 5.4.7   **CONCHAPELOPIA AGGREGATE CLAWS OF THE ANTERIOR PARAPODS**

Conchapelopia melanops

Arctopelopia barbitarsis

Rheopelopia maculipennis

Rheopelopia perda

Thiennemannimyia carnea

Rheopelopia ornata

Thiennemannimyia pseudocarnea

Thiennemannimyia morena

K

6 GENERAL ASPECTS OF THE SYSTEMATICS, BIOLOGY AND ECOLOGY OF THE TANYPODINAE

SYSTEMATICS

The evolutionary relationships between the genera of Tanypodinae have been studied by Fittkau (1962) and Saether (1977), with subsequent corrections and additions. In this book we only cover systematic relationships in the descriptions of single genera.

LIFE CYCLE

The life cycles found within the subfamily Tanypodinae are very diverse. Although most species can have two or three generations a year (depending on temperature and feeding), some species are univoltine, the latter having a diapause as young larvae or in the fourth instar. The time of emergence depends on the habitat: in *Anatopynia plumipes* in early spring, in *Telmatopelopia nemorum* in late spring and in *Clinotanypus nervosus* in early summer. Almost all species probably have a diapause in winter, but for many species this cannot be stated with certainty.

When describing separate species we usually give the emergence periods without mentioning fluctuations in the number of emergences within that period. As a rule they are more or less synchronised. However, the time of emergence depends on the availability of food and the temperature of the water bodies. Most species display a clear emergence peak in spring and less synchronisation in summer.

Oviposition and egg masses
Egg masses are often deposited against a firm substrate (often near the water's edge) or thrown off above open water (Lenz, 1936; Koreneva, 1959; Nolte, 1993). Few Tanypodinae are thought to deposit their eggs on moist soil (possibly *Krenopelopia, Natarsia* and *Telmatopelopia*). The shape of the egg mass and arrangement of the eggs differs considerably between genera. The egg mass may be globular, bale-shaped or club-shaped, and the egg row may form a spiral or the eggs may be arranged differently (Nolte, 1993). Koreneva (1959: 116) states that the females of different species deposit their eggs on their fifth or sixth day of their life, and then die.

Larvulae
In the **first instar** (the larvula) Tanypodinae are already recognisable by their retractile antennae, the singular kidney-shaped eye and the distinctively developed ligula (Thienemann and Zavrel, 1916; Mozley, 1979). The abdominal segments bear three long lateral setae. The ligula has two large teeth on either side of two very small teeth. The larvular mentum is similar to that of other Chironomidae, with a broad median tooth and three or four lateral teeth.

Further identification of first instar larvae is impossible. However, at this stage the first antennal segment in the Pentaneurini is already long (AR ± 3) and in the Macropelopiini is short (AR < 2). The larvulae are positively phototactic and at least those living in stagnant water are free swimming. For more information see Chapter 2 on general ecology (p. 10).

Second, third and fourth instars
In general the characters of the larvae remain constant from the second to the fourth instar.

There is no reliable rule for determining the instar of a larva. For identified larvae, the head sizes given for each species can be used. Identification of younger instars can be difficult because most keys make use of quantitative characters. This book contains keys for identifying species in the third and fourth instar.

The larvae of all Tanypodinae species are free-living. Some larvae may be found in a tube, but this is never a tube made by the larva itself.

Pupae
Tanypodinae pupae are also free-living. They look similar to Culicidae pupae because the thoracic horn has a plastron plate to take up atmospheric air. This plastron plate can be less developed in species of oxygen-rich running water. If the larvae are bottom dwellers the pupae can often be found between the vegetation or near the surface.

Adults
All Tanypodinae can disperse by flying. Larger species, such as *Anatopynia plumipes*, seem to fly more directly and can cover longer distances without drying out (see Chapter 2). Very little is known about swarming. Edwards (1929: 281) mentions that 'some Tanypodinae swarm rather high up'. Investigations of flying adult chironomids in Tilburg suggested that Tanypodinae are relatively scarce at greater distances from water. Nevertheless, some species are good colonisers (see under Dispersion below).

GENDER RATIO
Schleuter (1985) gives a gender ratio for Tanypodinae of 51% females. However she found more females for *Xenopelopia falcigera* (58–70%) and a variable ratio for *Zavrelimyia barbatipes* (26–83%). The author suspects that parthenogenesis is not exceptional in small water bodies (cf. Dettinger-Klemm, 2003: 242). Until now parthenogenesis has never been proved for any Tanypodinae species.

MICROHABITAT
The larvae are free-living, without a tube, and creep on or in the soil, on plants, stones, etc. They may be found in tubes, but they remain only temporarily in these tubes, which are always made by another species. Although the larvae are mostly found in the surface layer of the soil or on plants, Holzer (1980) found some larvae of *Thienemannimyia* (possibly including *Conchapelopia*) in the sandy/gravelly soil of the river Morava at a depth of 10–20 cm below the surface, and one larva deeper than 30 cm. In oxygen-poor soils the larvae probably only live in the surface layer, where they can benefit from the oxygen concentration in the water column and are less dependent on the oxygen in the sediment (Int Panis et al., 1995). Younger larvae in particular are frequently found in the water layer, before they settle in a suitable place, or when they have left the substrate, usually to find a better place or more food. In running water many larvae will be transported by drift.

WATER TYPE
Most Tanypodinae are inhabitants of stagnant water. The more plesiomorphic types (*Tanypus, Anatopynia, Psectrotanypus*) cannot live in fast running water. Some Pentaneurini species are better adapted to currents (*Rheopelopia, Nilotanypus*).

RESPIRATION
There is no frequent ventilation behaviour because the larvae do not live in a tube. Only the body fluid of *Macropelopia, Clinotanypus* and *Natarsia* species is red coloured; in some other genera it is slightly reddish. Many species require well aerated water and can only live in running water or oxygen-rich lakes and canals. It is probable that many larvae change their micro-

habitat when the bottom or even the water layer becomes poor in oxygen.

FEEDING

The food of Tanypodinae larvae has been investigated very thoroughly. Most investigators have studied gut contents, others have also made observations or conducted experiments. The extensive literature is often contradictory.

Some authors consider the larvae to be obligate predators. Both animal and vegetable matter is found in the gut, but several authors point out that the vegetable material may originate from sucked prey (Leathers, 1922; Wesenberg-Lund, 1943; Belyavskaya and Konstantinov, 1956). The last authors stated that smaller prey was swallowed whole, but larger prey only sucked out. Hildrew et al. (1985) noted that larvae of Plecoptera were always fragmented. A large variety of prey species have been observed in many Tanypodinae. Most frequently mentioned as prey are Cladocera and other small Crustacea, Oligochaeta and Chironomidae (Belyavskaya and Konstantinov, 1956; Armitage, 1968; Kajak and Dusoge, 1970). Loden (1974) observed larvae of Procladius, Labrundinia and Ablabesmyia actively attacking Oligochaeta and even trying to prevent the worm escaping. Thienemann and Zavrel (1916) mentioned that larvae of the Conchapelopia type also preyed on larvae of the same kind on a large scale. Konstantinov (1961) observed cannibalism only of smaller larvae, and only when the larvae face starvation.

It has been established that vegetable material is also eaten. Not only does the gut often contain only unicellular algae or plant remains, as in larvae of Tanypus (Roback, 1969) and Procladius (Armitage, 1968), larvae have also been directly observed swallowing unicellular algae (Thienemann and Zavrel, 1916). Mackey (1979) found filamentous algae in the gut of Ablabesmyia monilis. Biever (1971: 1167) ascertained that the larvae of Tanypus grodhausi were not predaceous during laboratory rearing. According to Hamilton (1965), Larsia acrocincta eats exclusively diatoms and detritus. Direct consumption of detritus has been reported by Hildrew et al. (1985) in Macropelopia, Trissopelopia and Zavrelimyia.

Accurate observations on feeding have been made by Izvekova (1980). She stated that larvae of Procladius ferrugineus were actively looking for animal food, but also consumed diatoms, blue-green and green algae, and even pollen. Non-moving particles were only consumed when they were right in front of their mouth parts; the larvae probably recognised them with the help of chemoreceptors. Only the ligula, paraligulae and labrum are used for swallowing such food, not the mandibles. Procladius ferrugineus larvae chase small chironomids that do not inhabit tubes. They penetrate into the tubes of larger chironomid larvae and drive them away, possibly to catch small animals that enter them. Mackey (1977) conducted an experiment in which he fed larvae of Ablabesmyia monilis with only detritus and another group with animal food (Stentor and Oligochaeta). The first group grew much more slowly than the second, and it was not clear if they could develop completely, but they continued to live and grow. Vodopich and Cowell (1984) found that third instar larvae of Procladius culiciformis feeding on algae and detritus were unable to moult to the fourth instar without animal food. Armitage (1968) concluded from his investigations that more algae and detritus were eaten if less animal food was available. Hildrew et al. (1985) showed that this behaviour (in Zavrelimyia barbatipes) in a brook in England was especially important in winter. Biever (1971) investigated the influence of food shortage and demonstrated that when rearing larvae of Tanypus grodhausi, development time was considerably prolonged at high densities, and in these conditions many young larvae were unable to complete development.

Many investigations of the feeding of Tanypodinae larvae are too fragmentary to prove clear differences in feeding between species. It seems unlikely that food specialisation has played an important role in the evolution of different species within one genus. In any case, the assessment by Moog (1995) that nearly all Pentaneurini live entirely on animal food is incorrect. This will be elaborated as far as possible in the descriptions of the separate genera and species.

In summary, it can be said that Tanypodinae of different groups feed on plant as well as animal material, and that most species can exploit both diets. Animal food plays an essential role in the development of most species.

PARASITISM AND PREDATION

According to Thienemann and Zavrel (1916), Tanypodinae larvae are largely free of parasitic Nematoda (*Mermis*) and Sporozoa are thought to be less common than in other Chironomidae. However Aagaard (1974) found parasitic nematodes in the genera *Procladius, Conchapelopia, Thienemannimyia* and *Ablabesmyia*. The adults displayed serious aberrations in genitalia and antennae, and their swarming behaviour seems to deviate from that of normal males. It is unknown if the mortality among larvae is also increased. Michiels and Spies (2002) also found some intersex specimens of *Conchapelopia hittmairorum* that were without doubt the result of parasitism by nematodes. Thienemann (1954: 305) mentions two species of Sporozoa in larvae of this subfamily.

Lenz (1936: 58) writes that fishes are the most important predators of the larvae, besides Hydracarina (especially *Hygrobathes*) and Hydrozoa. The adults are caught by flies, other insects (Schlee, 1977) and by birds (Tait Bowman, 1980). Most birds seem to select the largest prey and are probably more interested in *Chironomus* and other large chironomids than in Tanypodinae.

DISPERSAL

Dispersal of adults by air is not essentially different from dispersal in other subfamilies. We refer to Chapter 2, on the general ecology of the family. It is not clear if Tanypodinae disperse better or worse than other chironomids. Dispersal of larvae is covered briefly in chapter 2.2 on the life history of the Chironomidae. Tanypodinae are relatively frequently transported by drift, owing to the fact that the larvae move around freely and continually have to search for their prey. Moller Pillot (2003) concluded that in dynamic lowland brooks young larvae in particular are transported by drift and distances of many hundreds of metres are common. Drift is minimal when most larvae are in the fourth instar.

7 SYSTEMATICS, BIOLOGY AND ECOLOGY OF GENERA AND SPECIES

(A) DESCRIPTIONS

In this chapter the systematics and distribution of species will be treated briefly, and their biology and ecology more extensively. As a rule, we refer to the literature where necessary. For the ecology of the species in particular, we have drawn extensively from other sources, mainly data from investigations by many workers in the Netherlands who sent material and reports. The databases held by water authorities have not been investigated thoroughly.

This section contains written descriptions. For a numerical evaluation see the tables in section B and Moller Pillot and Buskens (1990).

Ablabesmyia   Johannsen, 1905

IDENTIFICATION OF SPECIES

In this book we follow the differences between the three European species published earlier (Moller Pillot, 1984). These differences are based on Laville (1971) and our own reared material of *A. monilis* and *A. phatta* (*A. monilis* has also been reared and described by Goddeeris, 1983: 135). Fittkau and Roback (1983) and Sergeeva (1998; 2004) give the reverse characteristics for the palpus of *A. monilis* and *A. longistyla*. Sergeeva (personal correspondence) wrote that she had mistakenly swapped the species; Fittkau could not check his material and wrote that making such a swap was possible. Moreover we found sometimes the basal part of the palpus of *A. monilis* divided into three segments, in one case only at one side.

We found some overlap between the exuviae of *A. monilis* and *A. longistyla*, as described by Langton (1991). Further, in Belorussian material some larvae and exuviae have sizes intermediate between *A. phatta* and *A. monilis*. However, adult males that could not be identified have never been found.

Our conclusion is that the original descriptions are most probably correct, but the possibility that the two species interbreed cannot be excluded. Therefore the palpus is not always reliable. A few specimens cannot be identified as a larva because of intermediate sizes. As proposed earlier in an unpublished key, we use the name *A. monilis* agg. for specimens belonging to *A. monilis* or *A. longistyla*, in contrast to *A. phatta*. Some workers will identify only to aggregate level because identification up to species level can be very time consuming, and because there are hardly any ecological differences between these species. In many cases two or even three species of *Ablabesmyia* are found living together.

FEEDING

The larvae of *Ablabesmyia* are predators, actively attacking Chironomidae, Oligochaeta and to a lesser extent also smaller swimming animals such as Cladocera (Konstantinov, 1961; Roback, 1969a; Loden, 1974; Mackey, 1979). However, dead prey, diatoms, filamentous algae and detritus are also eaten, especially when animal prey is scarce (Armitage, 1968). Mackey (1977) found that development was slow and possibly incomplete if the larvae received only detritus.

MICROHABITAT

The larvae live mainly among plants and on sediments containing organic material; they are nearly absent from fully mineral water bottoms, but found sometimes on wood or stones (Shilova, 1976; Bijlmakers, 1983; Klink, 1991; Verberk et al., 2005; Buskens: sand pits, unpublished). Bijlmakers (1983) found a slightly higher proportion on water bottoms with less organic material. In some places small differences were found between the microhabitat of different species of the genus, but such differences were not observed in other places. An exception could be A. *phatta*, which seems to live more among submerged vegetation (Steenbergen, 1993).

Ablabesmyia monilis   (Linnaeus, 1758)

DISTRIBUTION IN EUROPE AND THE NETHERLANDS

A. *monilis* has been recorded in all regions of Europe (Fittkau and Reiss, 1978). In the Netherlands the species seems to be less common than both other species of the genus, especially in the Holocene parts of the country (Nijboer and Verdonschot, 2001).

LIFE CYCLE

In late spring there is often a period without older *Ablabesmyia* larvae after the first emergence period because of the strong synchronisation. According to Goddeeris (1983) A. *monilis* has three generations a year in the fish ponds in Mirwart in the Belgian Ardennes. The larvae have a winter diapause in the second and third instar. In the Netherlands the first larvae in the fourth instar can be found in the second half of March. In most countries in Western Europe emergence first takes place from the end of April to the end of May; the second and third generations emerge mainly from July into September, and rarely to the middle of October (Mundie, 1957; unpublished Dutch data). In northern and mountain regions only one or two generations occur (Brundin, 1949; Laville and Giani, 1974 (A. *longistyla*)). The aberrant life cycle given by Mackey (1976) is probably based on incorrect identification (see Goddeeris, 1983: 25). As a result of local circumstances, the winter or summer generation can be very small or absent (Mundie, 1957; own unpublished data; compare Shilova, 1976: A. *phatta*).

FEEDING AND MICROHABITAT

See under the genus.

DENSITIES

The densities of larvae are usually low. In two pools near Oisterwijk, Bijlmakers (1983) rarely found more than 100 larvae/m². Brundin (1949) mentions much higher densities of Pentaneurini (often mainly *Ablabesmyia*) in Swedish lakes: as high as over 500 larvae/m² in June and more than 900 in October.

EGG DEPOSITION

The cylindrical egg masses contain 100 up to 400 eggs arranged in a spiral, which makes them more easily identifiable than those of most other Tanypodinae (Koreneva, 1959; Nolte, 1993). Koreneva found the egg masses attached to *Potamogeton* in the littoral zone.

pH

Leuven et al. (1987) found A. *monilis* at a pH from 3.8 to 7. In the Dutch province of North Holland the species was also not rare at a pH higher than 8 (Steenbergen, 1993). The larvae therefore seem to be indifferent to this factor.

WATER TYPE
Current

Ablabesmyia larvae are scarce in European streams (Lehmann, 1971; Lindegaard-Petersen, 1972; Braukmann, 1984). In the Netherlands the larvae of *A. monilis* are found in small numbers in lowland brooks and rarely in larger rivers.

Dimensions

All species of *Ablabesmyia* occur more often in larger stagnant water bodies more than 10 m wide (Moller Pillot and Buskens, 1990; Verdonschot et al., 1992; Steenbergen, 1993) and the larvae are scarce in narrow ditches. More larvae can be found in medium-sized lowland streams than in larger rivers. In lakes the densities are much higher in the littoral zone, less than 4 m deep (Mundie, 1957; Shilova, 1976). In deep water the larvae emerge much later than in the littoral zone (Brundin, 1949).

Permanence

Larvae of *Ablabesmyia* are rarely found in temporary water (Tourenq, 1975; Shilova, 1976; Schleuter, 1985; Delettre, 1989; Moller Pillot, 2003). In most cases eggs are deposited in spring, before desiccation. Winter generations are only found in a temporary pool if water was already present in autumn, when the last females fly (early October).

TROPHIC CONDITIONS AND SAPROBITY

In Bavaria Orendt (1993: 155) did not find the larvae of *A. monilis* in oligotrophic lakes, but commonly in mesotrophic and eutrophic lakes. They live mainly in β- and α-mesosaprobic water and are absent from polysaprobic water (Moller Pillot and Buskens, 1990; Moog, 1995). According to Dutch data they cannot survive anaerobic periods and can be compared in this respect with *Conchapelopia melanops*. Peters et al. (1988) found the larvae (of *A. monilis* and *A. longistyla*) in canalised lowland brooks only when the quality of the water was good.

SALINITY

In the Netherlands all species of the genus are rare in brackish water, especially if the chloride content exceeds 1000 mg/l (Krebs, 1981, 1984, 1990; Steenbergen, 1993). Steenbergen mentions the occurrence of *A. monilis* only in slightly brackish water (less than 1000 mg chloride/l). In Estonia, however, the larvae also live in water with much higher salinity, up to 6700 mg salt/l (Tölp, 1971). Such differences have been reported also in *Procladius*.

Ablabesmyia longistyla   Fittkau, 1962

SYSTEMATICS AND IDENTIFICATION
See under the genus.

DISTRIBUTION IN EUROPE AND THE NETHERLANDS
Fittkau and Reiss (1978) mention *A. longistyla* occurring throughout the whole of Europe, except Scandinavia. In the Netherlands it is the most common species of the genus almost everywhere in the country, especially in the Holocene areas.

BIOLOGY AND ECOLOGY
The life cycle and other biological characters seem to be the same as in *A. monilis*. Many authors (Fittkau, 1962; Verdonschot et al., 1992; Steenbergen, 1993; Moog, 1995) mention that *A. longistyla* is found in running water more often than *A. monilis*. On this basis some authors suppose that *A. longistyla* requires more oxygen and lives mainly in water of better quality (Fittkau, 1962:437; Wilson and Ruse, 2005). However, these differences have never

been found to be significant. In the Netherlands *A. longistyla* is the more common species in stagnant water and *A. monilis* and has been found no less frequently in streams. The ratio between the occurrence of both species seems to vary locally, but until now no ecological difference has been proved. Clear differences between both species have not been reported for pH (Buskens, 1987; Leuven, et al., 1987) or salinity.

Ablabesmyia phatta   (Egger, 1863)

DISTRIBUTION IN EUROPE AND THE NETHERLANDS
A. phatta is known throughout the whole of Europe, with the exception of the Mediterranean area (Fittkau and Reiss, 1978). In the Netherlands the species is widespread, with the possible exception of some parts of Zeeland and South Limburg.

LIFE CYCLE
The life cycle of *A. phatta* seems to be the same as *A. monilis*. However it is not yet clear if there are two or three generations. In Russia Shilova (1976) found only one generation in temporary water, but elsewhere two. In peat cuttings in Groote Peel (Netherlands) only the first generation was numerous (Werkgr Hydrobiol., 1993), elsewhere many pupae were caught in May, July and August.

FEEDING AND MICROHABITAT
See under the genus. More than the other *Ablabesmyia* species the larvae seem to live among submerged vegetation (Steenbergen, 1993).

DENSITY
Bijlmakers (1983) always found fewer than 100 larvae/m² in the Staalbergven near Oisterwijk.

pH
Buskens (1983) and Verstegen (1985) consider this species characteristic of acid moorland pools. However, the larvae are also fairly common in North Holland in water bodies with pH 8 or higher (Steenbergen 1993). Leuven et al. (1987) found the larvae of *A. phatta* in water at a pH of from 3.46 to 9.45.

WATER TYPE
Current
In contrast to the other species of the genus, *A. phatta* does not live in running water. There are just a few recorded occurrences of the species in very slow flowing lowland streams.

Dimensions
Shilova (1976) mentions that *A. phatta* is more common in small water bodies than *A. monilis*. In the Netherlands this is hardly ever the case.

Permanence
The larvae have been rarely found in temporary water (see under *A. monilis*).

TROPHIC CONDITIONS AND SAPROBITY
In Sweden Brundin (1949) found *A. phatta* especially in oligotrophic lakes. Fittkau (1962) supposes that the species is less euroxybiont than *A. monilis*. Steenbergen (1993) found that in the Dutch province of North Holland *A. phatta* occurred in water bodies with less phosphate and chlorophyll-a than water where *A. longistyla* was found.

SALINITY

Although *A. phatta* seems to be mainly a freshwater species, some records are known from water with a chloride content above 1000 mg/l (Krebs, 1984: 119; Steenbergen, 1993: 481).

Anatopynia plumipes (Fries, 1823)

SYSTEMATICS

A. plumipes is the only species of the genus in Europe. Identification of adults, pupae and larvae presents no problems. The genus has many plesiomorphic characters and is considered as a very primary type within the Tanypodinae (Fittkau, 1962). A separate tribe Anatopyniini has been created for this genus.

IDENTIFICATION OF ADULTS

The adult male is not keyed in Pinder (1978) because the genus is absent from the British isles.

DISTRIBUTION IN EUROPE AND THE NETHERLANDS

The species has been found throughout Middle, Northern and Eastern Europe, but is absent in the west and south, which can be explained by the fact that a cold period must be passed through to terminate diapause (see under *Life cycle*). The species occurs throughout the Netherlands, except in parts of Zeeland and on the Wadden Islands.

LIFE CYCLE

A. plumipes has only one generation a year. Pupae and adults were observed in the Netherlands from the middle of February until early April. Elsewhere in Europe the adults fly a little later; Fittkau (1962) gives April and May, and in the south of Belarus the pupae were locally numerous at the end of April 2003 (own observations). In the Netherlands the larvae are in the third and fourth instar in July. Shilova and Zelentsov (1972) have shown that the end of the diapause (in the fourth instar) is not influenced by day length, but by passing through a cold period.

FEEDING

Hardly anything is known about feeding. We found many juvenile Ostracoda in the gut of a second instar larva.

MICROHABITAT

The larvae usually live on top and in soft silty soils. Pupae are often found between vegetation.

SWARMING AND OVIPOSITION

Adult males have been observed swarming at a height of about two metres. The adults are able to fly in a straight line because of their large dimensions. We observed deposition of an egg mass on *Glyceria maxima* along the margin of a ditch. The egg masses are more or less globular, or slightly oblong, 6–8 mm long; they contain 300 to 700 eggs (see also Nolte, 1993).

DENSITY

In the Netherlands usually only a few larvae are found in a sample. During a limited investigation in a Frisian ditch in winter Beattie (1978) found 150 larvae/m'.

WATER TYPE

The larvae live in medium-sized stagnant water bodies such as ditches and small lakes, in meadows, marshes and bogs. In the province of North Holland, Steenbergen (1993) found the larvae most often in water bodies with little emergent vegetation. Owing to their uni-voltine life cycle they cannot survive in temporary water unless they can retire into the moist bottom.

SOIL

In the Netherlands the larvae occur most often in peaty regions, in fens as well as in bogs. They can also live elsewhere, if there is a silty soil. *A. plumipes* is not as scarce in sand as in clay.

pH

The larvae live in water with a pH from 4.5 to 9.

TROPHIC CONDITIONS AND SAPROBITY

The larvae are often found in water more or less affected by eutrophication, guanotrophy, etc. They can survive for a long time in water almost depleted of oxygen. It is not known if they have adaptations other than haemoglobin, for instance in their behaviour.

Often more mesotrophic conditions prevail and usually the phosphate content is low; Steenbergen (1993) gives a median orthophosphate content of 0.09 mg/l.

SALINITY

Although the species is almost only known in freshwater, in some cases larvae are found in slightly brackish water up to 1000 mg chloride/l or more (Steenbergen, 1993).

DISPERSAL

The adults are good flyers and can easily colonise new water bodies. In fact, the species has been found in isolated pools.

Apsectrotanypus trifascipennis   (Zetterstedt, 1838)

SYSTEMATICS

The genus *Apsectrotanypus* is probably most closely related to *Psectrotanypus* (Fittkau, 1962); in the older literature these are included in this genus. However the larva bears greater resemblance to *Macropelopia*.

EUROPEAN SPECIES

In Europe *A. trifascipennis* is the only species of the genus.

DISTRIBUTION IN EUROPE AND THE NETHERLANDS

A. trifascipennis occurs widely throughout Europe, but except in boreal regions it is confined to running water. In the Netherlands controlled data are known from the Pleistocene sandy regions and South Limburg.

LIFE CYCLE

In southern England Pinder (1983) found two or three generations a year. Lindegaard and Mortensen (1988) estimate two generations in a brook in Denmark. In both cases third and fourth instar larvae were present the whole year round because the generations overlap. Pupae were most numerous in May and from June to September. It should be noted that the spring generation in the Pinder brook was less numerous (owing to drift of larvae in winter), but in the Danish brook was more numerous. In the Netherlands pupae and exuviae have been collected in every month from the middle of April to September. Both there and in Denmark some larvae are still in the third instar in early April, and so emergence proceeds gradually over a longer period. It has been clearly established that the species has a diapause in winter.

FEEDING

The feeding of *A. trifascipennis* has not been investigated specifically. This species probably feeds in the same way as *Macropelopia*, *Psectrotanypus* and *Procladius*: the larvae will be mainly carnivorous but also feed on algae and detritus, especially the younger instars.

MICROHABITAT

The larvae are typical bottom dwellers. Pinder (1980, 1983) stated that the larvae developed numerously only in detritus rich sediments in summer. Tolkamp (1980: 189) found the larvae significantly more often in sandy soil with much organic material. There was a preference for bottoms with both coarse and fine detritus.

DENSITIES

The larvae occur as a rule in low densities. In autumn Pinder (1983) sometimes found more than 1000 larvae/m^2 (locally much more, where coarse organic material had been deposited). In the Netherlands such densities may occur in Limburg; in typical lowland brooks the numbers remain well below 100/m^2. Pinder (1983) stated that in his lowland brook the population was decimated after autumn floods. Success was probably largely dependent on the presence or disruption of organic detritus, which determined the suitability of the environment.

WATER TYPE

The larvae are cold stenothermic and live in the greater part of Europe only in fairly fast-flowing streams, but in the northern part of Scandinavia also in lakes (Brundin, 1949; Fittkau, 1962; Lehmann, 1971). Moog (1995) mentions the species also for Alpine lakes. In the Netherlands the larvae are scarce in slow-flowing lowland brooks and fairly numerous in more or less fast-flowing brooks. Verdonschot et al. (1992) and Verdonschot (2000a) unjustifiably do not mention the species for the latter category. Large streams are not an appropriate habitat, therefore Caspers (1980, 1991) does not mention the species for the river Rhine.In temporary upper courses of brooks only incidental short time settlement is possible (Moller Pillot, 2003).

pH

Moller Pillot (2003) found, that the larvae were not rare in the Roodloop at pH < 6. The larvae were absent in the very acid upper course, but it was not clear if pH or summer drought played a dominant role here. No further information is available about occurrence in acid streams. Orendt (1999) did not find the larvae at pH < 6 and considers *A. trifascipennis* to be extremely sensitive to acid.

TROPHIC CONDITIONS AND SAPROBITY

According to data incorporated in Moller Pillot and Buskens (1990) the larvae cannot survive prolonged anoxia, which is consistent with its frequent occurrence in fast-flowing brooks. Owing to their preference for organic rich sediments they can live in organically polluted water, mainly in winter, or in fast running brooks. Peters et al. (1988) even found *A. trifascipennis* only in organically polluted, not very slow-flowing brooks (these investigations were confined to canalised lowland brooks).

DISPERSION
Moller Pillot (2003) found the larvae relatively often in drift samples. The species probably disperses often by flying females.

Arctopelopia   Fittkau, 1962

SYSTEMATICS AND RELATIONSHIPS
The genus *Arctopelopia* belongs to the *Conchapelopia* aggregate (= *Thienemannimyia-*Reihe Fittkau 1962) in the Pentaneurini. These genera are easily identifiable as pupae, but are very difficult to identify in the larval stage.

THE EUROPEAN SPECIES
According to Fittkau (1962) and Murray and Fittkau (1989) three species are known from Europe. The identification of adults of the three species is possible using Fittkau (1962), taking note that the names in Figures 133 and 134 have been swapped. However, Pinder (1978) illustrates an aberrant specimen, possibly belonging to a fourth species. Fittkau and Roback (1983: fig. 5.5) illustrate a larva from Austria differing from *A. barbitarsis* and perhaps belonging to an undescribed species.

A further question is the remarkable ecological variation within *A. barbitarsis* (Zetterstedt, 1850; nec Goetghebuer, 1934). The larvae live in oligotrophic lakes in Scandinavia and the Alps (Fittkau, 1962). The possibility that pupae attributed to this species from Dutch lowland brooks might possibly belong to a new species cannot be excluded; their identity is supported by the fact that Goetghebuer (1923) also found this species (as *Tanypus nigroscutellatus*) in Postel (type locality) in the Belgian lowland along the Dutch border. *A. barbitarsis* and *A. nigroscutellatus* are synonyms according to Fittkau (1962), and the hypopygium illustrated by Goetghebuer (1923: 108) resembles *A. barbitarsis* (and definitely not the aberrant species of Pinder). However, the gonostyle illustrated by Goetghebuer still raises some doubts. It is highly possible that Fittkau did not check the type specimen and based his conclusions on Brundin (1949: 674), who only saw the description. No adult males have been found in the Netherlands.
Ablabesmyia barbitarsis Goetghebuer, 1936 must be another species, most probably within the genus *Zavrelimyia* (see Brundin, 1949: 674; Fittkau, 1962: 317).

DISTRIBUTION IN EUROPE AND THE NETHERLANDS
A. melanosoma (Goetghebuer, 1933) is an arctic species (Greenland, Scotland, possibly Scandinavia). The type locality of *A. griseipennis* (van der Wulp, 1858) is The Hague (Netherlands), but in the 20th century the species has been found only in boreal and Alpine regions and in England (Fittkau, 1962; Langton, 1984; Orendt, 1993; Ruse, 2002). *A. barbitarsis* has the same distribution, but is present also in the eastern and southern parts of the Netherlands in small lowland rivers (Reest, de Wijk, 1974, 1991, 1992; Lee, Gramsbergen, 2000; Rozep, Oisterwijk, 1993, 1997). The presence of *A. griseipennis* at Postel (near the Dutch border in Belgium) reported by Goetghebuer (1934) is incorrect (Fittkau, 1962: 205).

The description of the ecology will be confined to A. barbitarsis, unless otherwise stated.

LIFE CYCLE
In the Netherlands prepupae, pupae and exuviae of *A. barbitarsis* have been found at end of April and in early May. The specimen in Postel (Belgium, as *Tanypus nigroscutellatus*, see above) was caught in May. Although *A. griseipennis* has a one year cycle in Iceland (Lindegaard, 1992) a second generation cannot be excluded for the Dutch populations of *A. barbitarsis*.

FEEDING

Most probably the larvae feed mainly on animal prey, like the larvae of the related *Conchapelopia*. Some algae and detritus will also be consumed.

WATER TYPE

In boreal and Alpine regions *A. barbitarsis* is a species of oligotrophic lakes (Brundin, 1949; Fittkau, 1962). Langton (1984) mentions the species in a pool in southern England. In the Netherlands the three known localities are all small lowland streams flowing at less than 50 cm/sec. In England *A. griseipennis* is a typical inhabitant of acid lakes (Ruse, 2002) and in Bavaria it lives in oligotrophic lakes (Orendt, 1993). In the Netherlands it was found only in the 19th century.

SAPROBITY

According to Moog (1995), in Austria *A. barbitarsis* and *A. griseipennis* are more or less typical inhabitants of oligosaprobic water. These species are probably more tolerant in the European lowlands. So far, larvae of *A. barbitarsis* have been found in slightly polluted brooks and rivers without a stable oxygen regime.

Clinotanypus nervosus   (Meigen, 1818)

SYSTEMATICS AND RELATIONSHIP

Clinotanypus nervosus is the only European species of the tribe Coelotanypodini. Characteristic for the larvae of this tribe are the conical head capsule, the location of the row of teeth on the dorsomentum and the pointed papilla between the procerci.

IDENTIFICATION OF ADULTS

The male adult can be identified easily with all existing keys. The female imago has been described by Sergeeva (2000).

DISTRIBUTION IN EUROPE AND THE NETHERLANDS

C. nervosus lives in the whole of Europe except in montane regions. In the Netherlands the species seems to be absent from large parts of Zeeland, most Wadden Islands and parts of South Limburg.

LIFE CYCLE

In Sweden Brundin (1949) found adults only in June;. In the Netherlands prepupae, pupae and adults are found mainly in June and exclusively from early May until the middle of August. Third instar larvae are known from late summer, winter and spring; fourth instar larvae can be found all year round. Evidently there is only one generation a year and third and fourth instar larvae are present together owing to differences in developmental time or a diapause.

FEEDING AND LARVAL HABITS

The larvae probably creep slowly in the upper layers of the sediment. This can be concluded from the small eye, the high quantities of haemoglobin and the fact that they often live in very soft muddy soil.

Not much is known about feeding habits. In the United States Roback (1969a) investigated the guts of 24 larvae of *C. pinguis*. In 12 guts he found remains of Tubificidae, sometimes also Chironomidae and Diatomeae. Moog (1995) calls *C. nervosus* a pronounced predator. It seems probable that the larvae are specialised in feeding on Oligochaeta because they are regularly found in soils where other prey is scarce.

MICROHABITAT

The larvae are pronounced bottom inhabitants. Brundin (1949) mentions a preference for bottoms with coarse detritus. Occurrence on pure sand or clay is rarely mentioned in the literature and the larvae nearly always live on bottoms with a high organic silt content, where they are sometimes the only species of chironomid present. Rarely, a single larva may be found on a water plant or a stone.

DENSITIES

In the oligotrophic Lake Innaren in Sweden, Brundin (1949) found a mean of 440 larvae/m^2 (with a maximum of 1370) at suitable sites. According to Verdonschot et al. (1992) in the Netherlands the larvae are never present in high densities. Steenbergen (1993) also rarely found a large number of larvae in a sample.

OVIPOSITION

According to Koreneva (1959) an egg mass contains 300 to 400 eggs. It is not known where the egg masses are deposited.

WATER TYPE
Current

C. nervosus is particularly common in slow to very slow-flowing water and in ditches, but is also found in lakes and moorland pools. The larvae become more numerous in lowland brooks after canalisation, probably largely because of the accumulation of silt on the bottom. The larvae are absent from fast-flowing brooks and rivers (Lehmann, 1971; Braukmann, 1984).

Dimensions

Brundin (1949) reports this species in sheltered bays of the Swedish lakes. In the Netherlands the species appears to be scarce in large lakes and nearly absent from large rivers. Steenbergen (1993) found the larvae most often in water bodies with a width of less than 4 metres and a depth of less than 50 cm. Very narrow ditches seem to be less appropriate (possibly for oviposition).

Permanence

As a rule the species cannot complete its life cycle in temporary water bodies because these usually dry up during or after the flying period.

SOIL

As a typical inhabitant of mud, C. nervosus is most common on peaty soils. The larvae are also often found in samples of sandy soil, but the species is scarce in ditches on clay (among others Steenbergen, 1993). However we found regularly larvae in a slow flowing water course with a bottom of clay mixed with organic silt.

pH

There are some records of a single larva or pupa in very acid water down to pH 3.8 (Mossberg and Nyberg, 1980; Werkgroep Hydrobiologie, 1993). All other data concern pH values in the range 6.5–9 (Leuven et al., 1987; Steenbergen, 1993). The rare occurrence of the species in acid water is probably mainly related to food; besides the pH in the mud can be higher than in the water layer. Verdonschot et al. (1992) report finding of C. nervosus in acid environments, but not in oligotrophic stagnant water.

TROPHIC CONDITIONS AND SAPROBITY

In Sweden Brundin (1949) found C. nervosus sometimes numerously in oligotrophic water.

In the Netherlands oligotrophic water-bodies are probably usually too acid or the water bottom consists of bare sand. The preference for nutrient rich water in the Netherlands may be due primarily to the availability of food. The species is also common in mesotrophic ditches in peaty regions. According to many investigators *C. nervosus* is not an inhabitant of strongly polluted water. Peters et al. (1988) found an increase in numbers of larvae in canalised brooks as water quality improved. The larvae are also found scarcely in water with a high ammonia content (above 1 mg/l) (de Graaf, 1983; Steenbergen, 1993). This can be related to the fact that Tubificidae (probably often the most important food source) are generally more susceptible to ammonia than Chironomidae (Moller Pillot, 1971).

SALINITY
The species seems to avoid brackish water. There are, however, some records from oligohaline water with a chloride content of at least 1000 mg/l (Steenbergen, 1993).

DISPERSION
Immigration via the air, followed by egg deposition, is probably common even over distances of 500 metres or more (Moller Pillot, 2003).

Conchapelopia   Fittkau, 1957

SYSTEMATICS AND RELATIONSHIP
Fittkau (1962) considered the genus *Conchapelopia* (created by him) to be related to *Rheopelopia*, *Arctopelopia* and *Thienemannimyia* (the *Thienemannimyia*-Reihe, earlier known as *Pelopia* gr. *costalis*). Saether (1977: 49) argued that these genera do not form a monophyletic unit. In view of the great similarity of the larvae of these four genera, we maintain here the *Thienemannimyia*-Reihe sensu Fittkau as *Conchapelopia* aggregate. Within the genus *Conchapelopia*, *C. melanops* is possibly the only species in the Netherlands. The other species, therefore, will not be treated extensively here. For a recent survey of the genus and identification of adult males and exuviae, see Michiels and Spies (2002).

Conchapelopia melanops   (Wiedemann, 1818)

DISTRIBUTION IN EUROPE AND THE NETHERLANDS
Conchapelopia melanops is found throughout Europe. In the Netherlands the species is common and often numerous in the Pleistocene areas and in South Limburg; elsewhere only scattered records are known, especially in regions where flowing water is scarce. The species is probably absent from Zeeland and the Wadden Islands, with the exception of Texel.

LIFE CYCLE
In a cold brook in Central Jutland (Denmark) *C. melanops* had only one generation a year, with an extended emergence period in summer (Lindegaard and Mortensen, 1988). For the Fulda region (Germany) Lehmann (1971: 474) gives April to August as the flying period. Wilson (1977) sampled exuviae from April until October in the River Chew in England. In the Netherlands the species emerges from mid April until the middle of October. After immigration in spring many larvae are already in fourth instar in July, which suggests that as a rule there are two generations a year (see Moller Pillot, 2003 and others). The Dutch data indicate two emergence peaks: in the second half of May and the second half of August. However, pupae and exuviae are found frequently in all summer months, indicating that here also emergence periods are extended. The difference from the Danish investigation is

probably largely explained by the low water temperature of the Danish brook. Lindegaard and Mortensen (1988) reported a break in the presence of fourth instar larvae; there is not such clear break in the Netherlands. Part of the second generation may emerge only in the following spring and in winter part of the population is in third instar. There is no obvious evidence of a diapause.

FEEDING

The fourth instar larvae at least are predators, eating (mostly smaller) chironomids, Oligochaeta and to a lesser degree small Crustacea. Cannibalism has also been observed (Thienemann and Zavrel, 1916; Konstantinov, 1961). Kawecka and Kownacki (1974) often found diatoms, green and blue-green algae and undetermined organic matter in the gut of *C. pallidula*. They could not decide if this material came from the alimentary canal of the consumed animals or was directly picked up from the bottom. In winter we also found mainly green algae in the guts of the larvae. It is highly likely that algae form a normal part of the diet when little animal food is available, as has been stated for many other species of the subfamily (see the general ecology of the Tanypodinae in Chapter 2).

MICROHABITAT

According to Fittkau (1962) the larvae live mainly on and between plants, mosses and algae, and to a less extent on the bottom of brooks and lakes. They are present in silty soil only if it is well aerated. The claim by Sterba and Holzer (1977) that '*Thienemannimyia*' larvae live up to 50 cm deep in the soil of a brook with sand and gravel in fact relates to *C. melanops*. Lehmann (1971) refers mainly to the presence of the larvae on stones. Spänhoff et al. (2004) found the larvae in small numbers on dead branches in a German lowland brook. In the Netherlands the larvae often even live near the tops of grassy plants with floating leaves, such as *Glyceria* and *Sparganium*. In woodland brooks without vegetation Tolkamp (1980) found the larvae to be numerous on the bottom. They showed a marked preference for coarse organic material, such as dead leaves, but it is not yet clear whether they prefer dead leaves to living plants if both are available. There are indications that younger larvae in particular live on plants.

SWARMING

Dettinger-Klemm (unpublished) observed a swarm near a German brook at a height of 1.5 m (on 2 May 1998).

DENSITY

Lindegaard and Mortensen (1988: fig. 2) mention that young larvae can live in densities of more than 500/m^2, while in winter months densities still in excess of 200/m^2 were found. Davies and Hawkes (1981) found even higher densities in riffles incidentally and locally. The densities stated in the Netherlands are lower, but this may be a product of less accurate sampling.

WATER TYPE
Current and dimensions

In the international literature *C. melanops* is called a species of running water and lakes (Fittkau and Reiss, 1978; Reiss, 1984). Although Lehmann (1971) takes the species for an ubiquist in the Fulda region of Germany, Braukmann (1984) found the larvae most often in lowland brooks. *C. melanops* occurs hardly at all in the Alps (Fittkau, 1962). In the Swedish lakes *C. melanops* is widely distributed, but scarce (Brundin, 1949); in the lakes of Schleswig-Holstein, however, the larvae are numerous (see Fittkau, 1962). Records of larvae found in smaller pools are very rare, but sometimes many adults can emerge from pools in the floodplain of brooks and rivers (Schnabel, 1999).

Nearly all Dutch records refer to running water, from small spring brooks to large rivers. Densities in the hills of Southern Limburg are not obviously lower than in lowland brooks, but the species is scarce in the smallest brooks and in large rivers. The larvae are found in stagnant water in lakes and canals in the Netherlands, but as a rule the species seems to be absent from such water bodies. For instance, the species is completely absent from large sand pits (Buskens, unpublished). It is not clear why the Dutch lakes are so different from German lakes.

Permanence

The species establishes itself easily in (not very acid) temporary brooks, as soon as the watercourse is flowing again, but no egg depositing females fly after the middle of October (Moller Pillot, 2003).

pH

Conchapelopia was not observed in Norwegian lakes with a pH below 6 (Raddum and Saether, 1981). In the Netherlands a little acidification seems to be no problem, but the larvae are rarely found at pH lower than 5. In the Roodloop near Hilvarenbeek a pH a little below 5.0 seemed to be endured well for some time in spring 1994 (Moller Pillot, 2003). It is not clear to what extent the pH at the bottom was higher during that period and if this was important for the larvae.

TROPHIC CONDITIONS AND SAPROBITY

In the Netherlands larvae of *C. melanops* are usually found in eutrophic water. A problem is that oligotrophic brooks in this country are mostly acidified, so that both factors can be distinguished hardly. The presence in mountain brooks of the Fulda region of Germany (Fittkau, 1962; Lehmann, 1971) shows that the species is not confined to eutrophic situations. Braukmann (1984) found the larvae in Germany slightly more prevalent in silicate- than in carbonate-streams.

As to saprobity, Moller Pillot (1971) found by far the greatest densities in moderately polluted lowland brooks. The larvae were not absent where pollution was heavier if some oxygen was still present (see also Moller Pillot and Buskens, 1990). Also in investigations of somewhat faster running streams elsewhere in Europe *C. melanops* appears to endure organic pollution well, as long as oxygen supply is tolerable (Davies and Hawkes, 1981). In general, the species benefits from a higher level of nutrients, but the larvae are not absent from poorer environments. Where currents are weaker the oxygen content plays a more prominent role.

SALINITY

Although *C. melanops* has been recorded once as being present in brackish water at more than 1000 mg Cl/l (Steenbergen, 1993), it is a pronounced freshwater species which seems nearly always to be absent in brackish water.

DISPERSAL

C. melanops colonise the temporary part of the Roodloop nearly every year via egg-depositing females from elsewhere (Moller Pillot, 2003: 58). The distances covered in these cases are no more than 450 m. Distances of tens of kilometres can be a barrier, as appears from the absence of the species from the isle of Terschelling, where the small brooks form an acceptable, although not optimal, biotope. The larvae are present during the whole year in drift samples (Moller Pillot, 1971; 2003), but in the Roodloop the species remains in higer densities in winter than other chironomids.

Conchapelopia pallidula   (Meigen, 1818)

IDENTIFICATION
A complete key for all European species in the adult and pupal stages is given by Michiels and Spies (2002). The larva is indistinguishable from other species of the genus. However, Pankratova (1977) states incorrectly that in *C. pallidula* the basal part of the palpus maxillaris has two segments. The colour of the living larvae seems to be more greenish-yellow.

DISTRIBUTION IN EUROPE
C. pallidula is known to be present in large parts of Europe, including the Ardennes and Sauerland (Michiels and Spies, 2002) and in the Rhine downstream of Bonn (Caspers, 1991). Some older data may refer to *C. hittmairorum*. In the Netherlands the species is only found as exuviae, sampled in the large rivers.

LIFE CYCLE
Just like *C. melanops*, *C. pallidula* flies from the middle of April until the end of October.

ECOLOGY
Compared with *C. melanops*, *C. pallidula* inhabits somewhat faster-flowing water, rarely also in the littoral zones of lakes, such as Lake Constance and the Lunzer Untersee (see Michiels and Spies, 2002). Braukmann (1984) did not found the species in German lowland brooks. *C. pallidula* can also be found in the upper reaches in the lower mountains (Lehmann, 1971; Michiels, 2004).

Conchapelopia hittmairorum   Michiels and Spies, 2002

IDENTIFICATION
The larva of *C. hittmairorum* is described by Michiels and Spies (2002). The head capsule is somewhat smaller than in other species of the genus (740–810 µm), but is otherwise indistinguishable from other species of the genus. The authors give a key for adult males and pupal exuviae.

DISTRIBUTION IN EUROPE AND ECOLOGY
Although *C. hittmairorum* still seems to be confined to faster-flowing streams than *C. pallidula* (Michiels and Spies, 2002), occurrence in the Netherlands cannot be totally excluded. The nearest recorded site lies in the eastern part of Westfalia, about 130 km from the Dutch border.

Guttipelopia guttipennis   (Van der Wulp, 1861)

DISTRIBUTION IN EUROPE AND THE NETHERLANDS
G. guttipennis is a Holarctic species and most probably occurs nearly everywhere in Europe, except the far North and parts of Southern Europe (Fittkau and Reiss, 1978; Bilyj, 1988). In the Netherlands the species seems to be absent only locally, in Flevoland and in certain parts of Zeeland and North Holland. In North Holland the species is confined to areas which contain fresh water for a long time past: the dunes, the dune edge, the Vecht Lakes and the Gooi region (van der Hammen, 1992; Steenbergen, 1993).

LIFE CYCLE
In moderated parts of North America *G. guttipennis* flies from May until August; in Florida,

however, it flies the whole year round (Bilyj, 1988). In the Netherlands prepupae and pupae were sampled from May until July . Nothing is known about wintering and the number of generations.

MICROHABITAT

Bilyj (1988) writes that the larvae are mainly found between water plants. On water soldier (*Stratiotes aloides*) the larvae are dispersed and present in moderate densities, about 20 larvae/m² leaf surface (Higler, 1977). As far as is known, they creep slowly around on the plants (see Fittkau, 1962: 261), but are also found on the bottom of lakes. Brock (1985) found *Guttipelopia* larvae in several litter bags with decomposing leaf blades of *Nymphoides peltata*.

WATER TYPE

Current and dimensions

The larvae were not reported in running water by Lehmann (1971) or Braukmann (1984). In the Netherlands the larvae have been found mainly in pools, ditches and lakes; rarely also in slowly flowing water. Brundin (1949: 456) found this species only in some eutrophic and polyhumic lakes in Sweden. Other authors, for instance Kreuzer (1940), report its occurrence mainly in smaller stagnant water bodies, between *Sphagnum* as well as in eutrophic circumstances.

Permanence

Fittkau (1962) mentions *G. guttipennis* also for temporary water, which led Dettinger-Klemm (2003: 215) to rashly infer that the species can be a drought tolerant aestivator species. Dutch data do not give any indication for this, neither do Verdonschot et al. (1992) mention the species for temporary water.

SOIL

According to Steenbergen (1993) the larvae occur mainly on sand and peat and hardly on clay. The other Dutch data are consistent with this.

pH

Acidity does not seem to be directly important for larvae of *Guttipelopia*, because they are found between *Sphagnum* and at a pH above 8.

TROPHIC CONDITIONS AND SAPROBITY; OXYGEN

As mentioned above, the larvae occur in eutrophic as well in polyhumic oligotrophic water. Brundin (1949) points out that the larvae are apparently adapted to temporarily oxygen-poor conditions, but they are absent from water that is without oxygen for hours, at least in summer (Moller Pillot and Buskens, 1990).

SALINITY

G. guttipennis is not found in brackish water.

DISPERSAL

The fact that in the Dutch province of North Holland *G. guttipennis* did not colonise the former brackish water regions, and also has not been stated in Flevoland (see above), may indicate that the adults hardly disperse. However, it can also be connected with the predominantly clay soils in these regions.

Krenopelopia Fittkau, 1962

SYSTEMATICS AND RELATIONSHIP

The genus *Krenopelopia* was described by Fittkau (1962), who supposed that the genus had to be placed more or less separately within the Pentaneurini or near *Natarsia*. The larvae of these two genera can be confused unless the mandibles are compared, whereas the differences with *Natarsia* in the adult phase are very substantial. The generic relationships are not yet clear.

EUROPEAN SPECIES

At least two species live in Europe, and both are found in the Netherlands: *K. binotata* and *K. nigropunctata*. The species to which the larvae belong cannot yet be identified.

DISTRIBUTION IN EUROPE AND THE NETHERLANDS

The genus has been found in most parts of Europe (Fittkau and Reiss, 1978), but distribution in Mediterranean countries has hardly been investigated. Distribution in the Netherlands is insufficiently known because the larvae are semiterrestrial and are therefore not caught in usual sampling. In the western part of the country the genus will be scarce and possibly absent from brackish regions. Elsewhere the larvae may be found anywhere in the country at suitable locations.

LIFE CYCLE

According to Lehmann (1971) adults of *K. binotata* flying in the Fulda region (Germany) from May until August. Lindegaard et al. (1975) classified *K. binotata* under species with a slow seasonal cycle and summer adults. In a spring in Denmark they found adults only in August. In the Netherlands, pupae of *Krenopelopia* have been found in May, a prepupa as early as the end of April. *K. nigropunctata* emerged in September from a sample taken in August. Kouwets and Davids (1984) caught adults of this species in June, early July and the end of August. Two generations are probably the rule in the Netherlands. In all winter months up to the end of April larvae were in third and fourth instar.

MICROHABITAT

According to Lindegaard (1995) both *K. binotata* and *K. nigropunctata* live in springs, especially in the madicolous zone, the wet zone just above the water surface. He calls *K. binotata* a bryocolous species (living on or in moss). Lindegaard et al. (1975) found that larvae were much more numerous in water-covered moss than in the detritus zone in the Ravnkilde spring. The data from the Netherlands demonstrate that larvae of *Krenopelopia* can live almost anywhere with seepage or in permanent wet localities just above the water surface. On several occasions larvae have been found in wet grassland where no pools or other water bodies were present.

DENSITIES

Lindegaard et al. (1975) found densities varying from 254 larvae/m^2 in August to 5545/m^2 in May in the moss carpet of the Ravnkilde spring in Denmark. At most localities in the Netherlands the densities are very much lower.

OVIPOSITION

The eggs are probably deposited in wet soil or mosses.

WATER TYPE

Fittkau (1962) supposed that *K. binotata* is a typical inhabitant of springs, living only in boreal regions at lake shores. The larvae were not found in the streams themselves. They were later found at lake shores in Germany (see Lehmann, 1971). *K. nigropunctata* was reared once from a hygropetric biotope. In the Netherlands the larvae seem to be numerous in springs, but they have also been found widely in places with little or strong seepage: along the water edge in lowland brooks, ditches and ponds. In some cases, larvae have been found living in marshes or wet grassland. Adults of *K. nigropunctata* were reared by the authors from a reed swamp near Almere and from a quaking bog near Veenendaal.

SOIL
The larvae occur on very different soils; there appears to be no preference for clay, peat or sand.

pH
The larvae are found in more or less alkaline soils, but at least one larva has been found in an acid environment at pH < 5.0 (Moller Pillot, 2003).

Labrundinia longipalpis   (Goetghebuer, 1921)

SYSTEMATICS
L. longipalpis is the only representative of this genus in Europe.

DISTRIBUTION IN EUROPE AND THE NETHERLANDS
L. longipalpis is widely distributed, but scarce in Middle- and Western Europe and in Scandinavia (Fittkau and Reiss, 1978). In the Netherlands most records are from the fen-peat regions in Utrecht and North and South Holland, but the species has also been record-ed on the inner dune edge (Steenbergen, 1993), from north-west Overijssel and from North Brabant and Limburg.

LIFE CYCLE
In the Netherlands few larvae have been found from April to the end of August, and prepu-pae, pupae and exuviae in May, June and August. Brundin (1949) caught adults in Sweden in June and July. We can draw no conclusions about life cycle from these data.

FEEDING
Loden (1974) observed that an American species of *Labrundinia* caught Oligochaeta.

WATER TYPE
In Scandinavia *L. longipalpis* is characteristic of oligotrophic, mesohumic and polyhumic lakes (Brundin, 1949; Saether, 1979). In the Netherlands the larvae live mainly in the seep-age zone between the Pleistocene and Holocene parts of the country, usually in a more or less peaty environment. In one case the exuviae of one specimen was found in an olig-otrophic lake after restoration.

pH
In the Netherlands most water bodies containing *L. longipalpis* have a pH of 7 to 8.

Macropelopia   Thienemann 1916

SYSTEMATICS
The genus *Macropelopia* is closely related to *Apsectrotanypus* and *Psectrotanypus* (Fittkau, 1962; Murray and Fittkau, 1989). Today these genera form the tribe Macropelopiini in

Europe. Fittkau (1962) divided the genus in two groups: gr. *nebulosa* and gr. *notata*. In our region the first group is monotypical; *M. adaucta* belongs to gr. *notata*. The differences between the two groups are most pronounced in the pupal stage.

EUROPEAN SPECIES
About ten species are found in Europe (Fittkau and Roback, 1983), but only three of them are found in the Netherlands and adjacent lowlands. These three species will be described separately.

IDENTIFICATION OF ADULTS
The male adults can be identified using Fittkau (1962). This author gives some indications for identifying the females according to colour and wing markings (p. 116).

FEEDING
Gouin (1959) describes the behaviour and the structure of the larval head of this predaceous larva. Despite its usually slow movements, it can project its body forward with a sudden movement to implant its mandibles into the body of its victim. Investigations by Hildrew et al. (1985) showed that the larvae of *M. adaucta* in an English brook were mainly carnivorous and ate Chironomidae, Plecoptera and Copepoda. Much detritus was also eaten, especially in winter.

Macropelopia adaucta   Kieffer in Thienemann & Kieffer, 1916

Macropelopia goetghebueri Edwards, 1929; Fittkau, 1962; Moller Pillot, 1984

DISTRIBUTION IN EUROPE AND THE NETHERLANDS
M. adaucta occurs throughout Europe, with the exception of the Mediterranean area (Fittkau and Reiss, 1978), but no records are known from West and Central France (Serra-Tosio and Laville, 1991). Edwards (1929) declares the species to be common in England in peaty districts. In the Netherlands *M. adaucta* is almost confined to the Pleistocene sandy regions in the east and south of the country; there is one record from a small dune stream near Schoorl (near the west coast of the Netherlands).

LIFE CYCLE
According to Fitkau (1962) and Lehmann (1971) *M. adaucta* flies in the Fulda region of Germany from April until June. Hildrew et al. (1985) found predominantly second instar larvae in June, third instar in July and fourth instar in August, probably belonging to a second generation in late summer. In the Pyrenees there is only one generation a year (Laville and Giani, 1974). In the Netherlands pupae have been found from the end of March until the end of October. There are two, possible sometimes three, generations a year. The data indicate a winter diapause.

MICROHABITAT
The larvae are typical bottom dwellers. According to Kuiper and Gardeniers (1998) the species is present more often when the bottom is silty. The majority of larvae in an English stream with a stony bottom lived between dead leaves and at places with stones and leaves and were scarce on open stony stream beds (Hildrew et al., 1985).

DENSITIES
Hildrew et al. (1985) found a maximum density of about 300 larvae/m² in summer; in winter and spring the larvae were very scarce.

WATER TYPE

M. adaucta lives in both stagnant and flowing water. Fittkau (1962: 127) mentions a preference for boggy regions and lists records of samples from marshes and springs. Braukmann
(1984) found the species to be scarce in German lowland and mountain streams. In England
Ruse (2002) caught the larvae in acidlakes. Laville and Giani (1974) found the larvae in a lake
at a height of 2285 m in the Pyrenees. In the Netherlands there are many records from acid
bogs and peat cuttings in addition to the main locations in (often temporary) upper courses
of lowland brooks. In summer Moller Pillot (2003) found some larvae further downstream,
in the permanent part of the Roodloop. Also in spring the larvae can be present in the lower
courses of lowland brooks. It is not clear, to what extent drift in winter plays a role.

FEEDING

See under the genus.

pH

M. adaucta is often recorded as a species of acid conditions (Brundin, 1949; Fittkau, 1962;
Raddum and Saether, 1981; Hildrew et al., 1985; Ruse, 2002). Orendt (1999) found the larvae in German streams at pH values from 3.2 to 6.2. Hall (1951) mentions values between
6.1 and 7.0. In the Netherlands the species has been found most numerously where the pH
is lower than 6.0 (Moller Pillot and Buskens, 1990). Kuiper and Gardeniers (1998) mention
values from 3.8 to 7.5.

TROPHIC CONDITIONS AND SAPROBITY

Although the larvae of *M. adaucta* are found mainly in acid water, they seem to tolerate high
nutrient loads or saprobity not to a lesser extent than *M. nebulosa* (see under this species).
Prepupae and pupae are frequently found in polluted streams (unpublished data), but it is
possible that these larvae are carried down from the less nutrient-rich upper courses.

DISPERSAL

In the Roodloop brook, in the Dutch province of Noord-Brabant, the larvae of *M. adaucta*
are rather numerous in the summer months in the permanent stretch, which flows very
slowly (Moller Pillot, 2003). In winter and early spring when the current is stronger the larvae are scarce in this part of the stream. In early spring almost the whole population lives
further upstream, in the temporary, slow-flowing stretches and stagnant waters. The
author therefore includes *M. adaucta* among the habitat-shifting species. In winter transfer
by drift was important; recolonisation of the upper parts in late summer could be mainly by
flight because in most summers there are no moist places upstream. Hildrew et al. (1985)
also found hardly any larvae in winter and early spring along a permanent stretch of an acid
stream in England. The habitat-shifting strategy is probably fairly common in this species.

Macropelopia nebulosa   (Meigen, 1804)

DISTRIBUTION IN EUROPE AND THE NETHERLANDS

M. nebulosa has been found across the whole of Europe (Fittkau and Reiss, 1978). The
species is widespread throughout the Netherlands, but it can be scarce in regions without
running water (Moller Pillot and Buskens, 1990).

LIFE CYCLE

Lindegaard and Mortensen (1988) found only one generation a year in a cold Danish brook,
emerging in spring. According to Fittkau (1962) there are one or two generations a year,
depending on the region. Moller Pillot (2003) found a possible third generation in the

Roodloop. Pupae can be found in the Netherlands from March to early October; third and fourth instar larvae are present throughout the year, and third instar larvae are absent only at the end of April. Lindegaard and Mortensen (1988) found second instar larvae until January. The data indicate a winter diapause.

DENSITIES

Lindegaard and Mortensen (1988) estimated the mean density in a Danish brook to be 59 larvae/m². In the Netherlands much higher densities have been frequently reported (several hundreds per square metre), but exact data have not been published. Very different densities can be found within a short distance.

WATER TYPE

Fittkau (1962) includes *M. nebulosa* among the moderately cold stenothermic species, living in running water and also in lakes and ponds in northern and mountain areas. In the Fulda region of Germany the larvae inhabit rivers from the spring regions down to the lower stretches (Lehmann, 1971). Braukmann (1984) found the species in lowland and mountain brooks. In the Netherlands the larvae are most numerous in lowland brooks, in the temporary upper courses as well as in the permanent lower stretches. In South Limburg the larvae are present in helocrene springs, but absent in the fast-flowing stretches on the slopes of the hills. In the whole country the larvae can be found in some sand pits, lakes, pools and ditches, but here they are scarce (Krebs, 1981; Moller Pillot and Buskens, 1990; Steenbergen, 1993). The species can be numerous only in some dune lakes (Schmale, 1999).

MICROHABITAT

The larvae are typical bottom dwellers. Lindegaard-Petersen (1972) considers the larvae to be typical for places with abundant detritus and weak currents, part of the *Micropsectra praecox*-community (Marlier, 1951). Tolkamp (1980) found the larvae of *M. nebulosa* mainly in mixtures of sand and detritus, but relatively few in bare sand or detritus alone. According to Lindegaard-Petersen (ibid.) pupation occurs in the upper mud layer.

pH

Kuiper and Gardeniers (1998) found larvae of *M. nebulosa* at pHs in the range 4.5–8.5, on average at a slightly higher pH than *M. adaucta*. Moller Pillot (2003) stated that, in contrast to the latter species, *M. nebulosa* is very scarce at pHs lower than 4.5.

TROPHIC CONDITIONS AND SAPROBITY

From studies in the middle and lower courses of lowland brooks, Moller Pillot (1971) allocates *Macropelopia* to the Hirudinea group, which is characteristic of moderately polluted water. The larvae of *M. nebulosa* were mainly numerous in the presence of organic pollution. Experiments by Walshe (1948) indicate that the larvae survive anaerobic conditions for only a short time and are almost never found in anaerobic water (see also Moller Pillot and Buskens, 1990: 10). In stagnant water the larvae live almost exclusively in water with a more or less stable oxygen regime.

SALINITY

The larvae are typical inhabitants of freshwater, but sometimes they have been found in oligohaline water with a chloride content between 500 and 1000 mg chloride/l (Krebs, 1981; Steenbergen, 1993).

TOXIC CHEMICALS

Rasmussen and Lindegaard (1988) state that the larvae are very tolerant of iron and found the larvae in water containingup to 26 mg Fe^{2+}/l.

DISPERSAL

Larvae of *M. nebulosa* are transported by drift, as mentioned for *M. adaucta* (see above). However, in winter *M. nebulosa* survives better in flowing water. The moderately acid stretches of the Roodloop were easily colonised by flying females after desiccation (Moller Pillot, 2003).

Macropelopia notata (Meigen, 1818)

DISTRIBUTION IN EUROPE AND THE NETHERLANDS

The range of *M. notata* covers most of Europe where springs can be found. In the Netherlands the species seems to be almost confined to the eastern part of the country and will be very rare, except in South Limburg. Besides we reared this species from a reed marsh in Flevoland.

LIFE CYCLE

According to Fittkau (1962: 124) this species can emerge from April until October and have one or two generations a year (dependent of the habitat), emerging in spring and late summer. In the Netherlands pupae have been found in May, July and September.

DENSITIES

Very little is known about larval densities. Lindegaard et al. (1975) give a mean number of 180 larvae/m² in a Danish spring, and neither is the species numerous in the Fulda region of Germany (Lehmann, 1971). In the Netherlands only very few specimens have been identified.

WATER TYPE

Fittkau (1962: 124) mentions *M. notata* as a typical inhabitant of springs, both limnocrenes and helocrenes. According to Fittkau and Reiss (1978) and Lindegaard (1995: 119) the larvae can also live in spring brooks. The Dutch sites are helocrene springs and a reed marsh in Flevoland fed by seepage water.

pH

Orendt (1999) found *M. notata* in German streams at pH values of 3.9 and 6.5. In the Netherlands the species seems to be absent in acid water; in South Limburg the species has been found in more calcareous springs, similar to those in Denmark, with pH > 7.5 (Lindegaard et al., 1975).

DISPERSAL

The population found in Flevoland near Almere proves that adult females can fly several tens of kilometres to deposit their eggs.

Monopelopia tenuicalcar (Kieffer, 1918)

Ablabesmyia brevitibialis (Goetghebuer): Brundin, 1949: 675.

SYSTEMATICS AND RELATIONSHIP

M. tenuicalcar is the only species in Europe belonging to the genus *Monopelopia*. This genus belongs to the Pentaneurini and is probably most closely related to *Xenopelopia* (Murray and Fittkau, 1989), although the female genitalia do not indicate a strong relationship (Saether, 1977: 49).

IDENTIFICATION OF ADULTS
The adult male cannot be identified using Pinder (1978) because of the distinct brown bands on the abdominal tergites. An important character is the absence of the R_{2+3} on the wing.

DISTRIBUTION IN EUROPE AND THE NETHERLANDS
Netherlands the species seems to be absent from large parts of the province of Zeeland (see Krebs, 1981), from a number of polders with clay bottom and from most of the Wadden Islands.

LIFE CYCLE
Nearly all larvae sampled in October and November were in the third instar and had very little food in the gut (van Geloven and Thielen, unpublished). Larvae found in winter are mainly third instar, which indicates a diapause in autumn and winter. Although in the Netherlands the first adults were caught in April, most prepupae and pupae are found in May and others in July, August and September. Development appears to begin slowly in spring. Data on other Pentaneurini (see for instance *Xenopelopia*) suggest that in the Netherlands the species has two or three generations. For Central Europe Fittkau (1962) gives a flying period of May to August (scarcely in April and September). According to Brundin (1949) in Sweden the species flies in June and July.

LARVAL BEHAVIOUR AND FEEDING
The larvae creep around actively, mainly on plants near the water surface (see under Microhabitat). Detritus and green algae have been found in the gut, but the food of this species has been inadequately investigated. Most probably the larvae live on vegetable material an only incidentally some small animals (see the general ecology of the Tanypodinae, Chapter 6).

MICROHABITAT
The larvae are nearly always observed on plants and rarely on the bottom, although they may move to the bottom during periods of frost. In contrast to *Xenopelopia*, by far the greatest numbers are found near the water surface, for instance on *Lemna* and *Azolla* (van Geloven and Thielen, unpublished; confer Kreuzer, 1940) or *Stratiotes* (Higler, 1977). The larvae have also been found on *Nuphar, Ceratophyllum, Phragmites*, etc.

DENSITIES
M. tenuicalcar usually occurs in very low numbers, but larger numbers are possible. For instance, van Geloven and Thielen (unpublished) found a maximum density of 145 larvae in an area of 0.6 m^2 in a sample of *Lemna minor* at the surface of a peaty ditch in Ouderkerk aan de IJssel, on 26 October 1987. Higler (1977) found the larvae in *Stratiotes* fields very locally and in very different densities.

WATER TYPE
Current
M. tenuicalcar is nearly absent from running water.

Dimensions
The recorded locations of this species in the literature are restricted largely to peat pools and the marshy littoral zone of lakes. Most Dutch records concern ditches from 2.5 to 8 m wide. In water bodies wider than 10 m, the larvae are found among vegetation containing *Stratiotes, Sphagnum cuspidatum*, etc., and rarely in more open vegetation.

Permanence

Incidental presence of larvae in temporary pools is possible (Dettinger-Klemm, 2003), although the larvae do not survive complete desiccation.

SOIL

According to Kreuzer (1940), Brundin (1947) and Fittkau (1962) *Monopelopia tenuicalcar* is a typical inhabitant of peaty water bodies. In the Netherlands the species is only numerous in fen-peat regions and probably sometimes in bogs. We also found the species in Denmark widely distributed in bog pits and bog depressions (unpublished). Records on sand always (?) concern water bodies with more or less dense vegetation. The species is rare on clay (Steenbergen, 1993; Nijboer and Verdonschot, unpublished).

pH

Considering their frequent occurrence in bogs, the larvae endure a low pH very well. The fact that they are scarce in poorly buffered waters in the Netherlands (Buskens, 1983; Leuven et al., 1987; van Kleef, unpublished) cannot be caused by the low pH. In ditches in North Holland the larvae are also common at pH > 8.0 (Steenbergen, 1993), leading us to conclude that the larvae are insensitive to pH. The presence of enough food in the specific microhabitat is probably most important.

TROPHIC CONDITIONS AND SAPROBITY

According to Fittkau (1962) *M. tenuicalcar* rarely inhabits eutrophic water, but in the Netherlands this appears not to be true. Brundin (1949: 675) supposed correctly that the species is eurytrophic. Our investigations show that the larvae can endure a very low oxygen content for a long time, possibly because most larvae live near the water surface. They can be numerous in hypertrophic or saprobic ditches, and can also occur in very clean water, in poorly buffered waters and bogs. In such cases they seem to occur in a decomposing environment. Steenbergen (1993) found significantly fewer larvae in water with higher ammonium content (gradually diminishing from 0.1 mg/l). The key factors for the occurrence of this species are as yet not clear.

SALINITY

There are no records from water with a chloride content above 1000 mg/l.

Natarsia   Fittkau, 1962

Pelopia gr. *fulva*: Zavrel and Thienemann, 1921: 720–723, fig. 35–39.

SYSTEMATICS AND RELATIONSHIP

The genus was established by Fittkau (1962) and placed between the Macropelopiini and Pentaneurini. Saether (1977) classified the genus within the Macropelopiini on account of the female genitalia. It is currently considered to be a separate tribe Natarsiini. The antenna of the female has 15 segments, as in *Anatopynia* and *Macropelopia*, in contrast to the true Pentaneurini. The larva resembles the Macropelopiini by the relatively short head and red colour. Although there is some debate about the name *N. punctata* (Fabricius), Spies and Saether (2004) assert that it remains valid.

EUROPEAN SPECIES

In Europe there are two species: *Natarsia punctata* (Fabricius) 1805 and *Natarsia nugax* (Walker) 1856.

IDENTIFICATION OF ADULTS AND PUPAE

The adults of the two species can be identified easily: the male and female of N. punctata have dark markings on the wings, which are absent in N. nugax. Further differences are minimal (the swelling of the gonostyle in N. punctata shown in fig. 18C in Pinder (1978) is a little exaggerated). The pupae of N. punctata can be identified before emergence by the wing markings. N. fulva Kieffer, 1918 is therefore most probably the same species (cf. Fittkau, 1962: 155, 162), given that Fittkau saw no difference between the pupa of Kieffer and that of N. punctata. According to Langton (1991) the pupa of N. punctata has short lateral setae on segments VIII and IX, whereas that of N. nugax has long filaments. We hesitate to use this character, because the specimen of Langton has not been reared and our ? N. nugax exuviae have slightly shorter filaments than shown in Langton's plate 17c. Moreover, both types can be found at the same location.

DISTRIBUTION IN EUROPE AND THE NETHERLANDS

N. punctata probably lives throughout Europe. N. nugax has been recorded in West and Central Europe (Fittkau and Reiss, 1978; Ashe and Cranston, 1990). The genus has not yet been found everywhere in the Netherlands and may be absent from some clay regions; most sites are on the Pleistocene (see Moller Pillot and Buskens, 1990) and in the dunes along the coast (see Steenbergen, 1993). Occurrence is often underestimated because many larvae live at the water's edge or above the water level, but N. punctata in particular can be present everywhere, except possibly in brackish water; N. nugax has not yet become established as an adult (see also under Identification).

LIFE CYCLE

In the Netherlands pupae have been found from the end of April to June. Larvae seem to be rare in July and have never been found in August. However, adults of N. punctata could be reared in August from larvae caught in July, and Goetghebuer (1911) mentioned adults in August near Ghent in Belgium. Without doubt there can be two generations a year; third and fourth instar larvae are already more common in September. Adults have been found in Swedish lakes by Brundin (1949: 687, as Macropelopia punctata) in June and early July. These would have been monocyclic populations.

FEEDING

Little is known about feeding in the genus Natarsia. In the guts of larvae of an American species, Roback (1969a) nearly always found diatoms but only a few small Crustacea and Chironomidae. Diatoms and Ostracoda have been found in the guts of Dutch specimens.

MICROHABITAT

The larvae live on sand and organic material on the bottom of small streams and ditches, but also along several types of water above the water level. See also under Water type.

DENSITIES

The larvae are usually found in very low densities. More than 100 larvae/m² was an exception in our investigations, and such densities were hardly ever encountered in ditches of very slow-flowing or standing water.

OVIPOSITION

Because the larvae are often found above the water's edge and in quaking bogs, sometimes far from open water, the eggs are probably deposited in wet soil or mosses. This could explain why young larvae are rarely found.

WATER TYPE

In most of the literature *N. punctata* is described a species of springs and streams (Lehmann, 1971; Fittkau and Reiss, 1978; Lindegaard, 1995). In Austria and Sweden the larvae also live in lakes (Fittkau, 1962; Brundin, 1949). Other authors found the larvae in pools and ditches (see Fittkau, 1962: 161; Langton, 1991). In the Netherlands most data are on small lowland streams, but Verdonschot et al. (1992) mention *Natarsia* as a characteristic inhabitant of shallow, oligotrophic water bodies. Further investigations have revealed that the larvae often live above water level and sometimes also in wet grassland with seepage. In the Grote Peel bog remnant, *N. punctata* was found only in quaking bogs, in *Sphagnum* as well as in tussocks of *Juncus effusus* (Werkgr. Hydrobiol., 1993); evidently the species is not confined to just one or two water types. Langton (1991) calls *N. nugax* a species of streams, whereas in the Netherlands pupae and exuviae with filaments on segments VIII and IX (see under identification) have also been found in stagnant water.

SOIL

The larvae seem to be rare on clay, but they can live on clay substrates where water is more or less permanent, for example at the base of a dike.

pH

The larvae seem to be more or less unaffected by pH because they can live in *Sphagnum* and also in water with a pH > 8.

TROPHIC CONDITIONS AND SAPROBITY

According to most of the literature, *Natarsia* larvae are typical inhabitants of water bodies of good quality (see Moller Pillot and Buskens, 1990). Verdonschot et al. (1992) even mention oligotrophic waters. However, Zavrel and Thienemann (1921) found them also in polluted pools and ditches, where they are probably sustained by seepage of ground water or live mainly along the banks.

SALINITY

Larvae are not found in brackish water.

DISPERSAL

Moller Pillot (2003) found no larvae in drift samples, but they are not rare in the larger lowland streams in Noord-Brabant. In such cases it is not clear if the genus is a permanent inhabitant or an immigrant by drift.

Nilotanypus dubius   (Meigen, 1804)

DISTRIBUTION IN EUROPE AND THE NETHERLANDS

N. dubius occurs throughout Europe, but is almost confined to montane areas (Fittkau and Reiss, 1978). The species has been found quite near to the Dutch border in Belgium and Sauerland (Fittkau, 1962). *N. dubius* is probably not a permanent inhabitant in the Netherlands. Only one larva has been found in the floodplain of the river Waal in April 1994, after extremely high water during the preceding winter (Klink and Moller Pillot, 1996).

DESCRIPTION OF THE LARVA

See Kownacki and Kownacka (1968).

LIFE CYCLE

Pupae, exuviae and adults have been found in June, August and September (Fittkau, 1962; Kownacki and Kownacka, 1968).

WATER TYPE

The species is characteristic of small mountain streams (Fittkau, 1962; Braukmann, 1984). In Germany it has also been found further downstream, for instance in the Black Forest (Michiels, 2004). However, Caspers (1980, 1991) did not find it in the Rhine. The larva caught in the Waal in the Netherlands (Klink and Moller Pillot, 1996) was certainly brought down by the extremely high river water in the winter of 1994.

Paramerina cingulata   (Walker, 1856)

SYSTEMATICS AND NOMENCLATURE

The larvae of *Paramerina* resemble those of *Zavrelimyia* and *Schineriella* and are probably closely related to these genera (Murray and Fittkau, 1989). In the Netherlands *P. cingulata* is the only representative of the genus. The use of the species name *P. pygmaea* (van der Wulp, 1974) by Moller Pillot and Beuk (2002: 125) is invalid (Spies and Saether, 2004).

DISTRIBUTION IN EUROPE AND THE NETHERLANDS

P. cingulata is known to occur in the whole of Europe except the Mediterranean region (Fittkau and Reiss, 1978). It is not clear whether the exuviae (Pe 1 Langton) found in southern Spain by Casas and Vilchez-Quero (1989) come from this or a related species (see Langton, 1991). The species has been found throughout the Netherlands, with the exception of Zeeland, South Limburg, the Wadden Islands and the Flevopolders. Although the larvae are fairly common (Nijboer and Verdonschot, 2001), they are found only incidentally and often in low numbers. On clay soils suitable water bodies may be scarce, even if the water is not brackish. The larvae will be present in most EIS plots.

LIFE CYCLE

Prepupae and pupae are found in the Netherlands from the end of March until September. As in *Zavrelimyia nubila*, the available data contain three indistinct peaks, in April, July and September, which indicates probably three generations a year. There are few data from the winter months, but some larvae caught in December and January were in the second and third instar.

FEEDING

Observations about feeding habits have not been published. Sometimes juvenile chironomids have been found in the gut.

MICROHABITAT

Brundin (1949) mentions the occurrence on organic as well as mineral soil in Swedish lakes. In the Netherlands the larvae appear to live often on water plants, even on floating duckweed. Higler (1977), however, does not mention the species from *Stratiotes*.

WATER TYPE

Current

P. cingulata is rare in fast-flowing streams, although the larvae can be found incidentally where the current is slower (Fittkau, 1962; Lehmann, 1971; Braukmann, 1984; Orendt, 2002). In Dutch lowland streams the species lives in low numbers, but is widespread.

Dimensions
Although the larvae occur fairly commonly in the bank zone of Swedish lakes (Brundin, 1949), in the Netherlands only a few records are known of the species in large water bodies, with the exception of lakes in fen-peat regions. The larvae are very rare in sand pits; Buskens (unpublished) found them only once between helophytes. Langton (1991) also reports the presence of this species in northern and montane lakes, but not in large lakes elsewhere. The larvae are virtually absent from the North German lakes (Fittkau, 1962). The Dutch data refer mainly to the bank zone of ponds, peat cuttings, peat lakes and floodplain pools; there are few observations in narrow ditches. This all suggests that *P. cingulata* occurs mainly in medium-sized and small water bodies.

Permanence
The larvae seem to be absent from temporary water.

SOIL
A relatively high proportion of records are from sand or peat (Steenbergen, 1993, etc.). In clay regions the species occurs locally, if water of a suitable quality is present.

pH
The larvae are absent from very acid water and are rare at pHs lower than 6. Many records are from seepage regions, for instance near the boundary between Pleistocene and Holocene landscapes. In the province of North Holland the species is characteristic of the dune margin and the Vecht Lakes region (van der Hammen, 1992; Steenbergen, 1993). Many records from brooks can also be interpreted in this way, but the species is definitely not an absolute seepage indicator.

TROPHIC CONDITIONS AND SAPROBITY
Apart from conditions with seepage or currents, the water is never hypertrophic or polluted. Streams in which *P. cingulata* lives permanently always have a fair to good water quality. Larvae living near the water surface on plants are less dependent on the oxygen content.

SALINITY
The larvae are rare at a chloride content higher than 300 mg/l.

Paramerina divisa   (Walker, 1856)

IDENTIFICATION
The larvae of *P. divisa* are easily recognised because all claws of the posterior parapods are simple. For a description, see Laville (1971). All claws of the posterior parapods are yellow. Moller Pillot (1984) writes incorrectly that the second antennal segment is light coloured.

DISTRIBUTION IN EUROPE
P. divisa is not known from the Netherlands, but has been found in the surrounding countries in mountainous regions, and in Scandinavia also in lakes (Fittkau, 1962; Serra-Tosio and Laville, 1991). Nijboer and Verdonschot (2001) mention a specimen from the Dutch province of North Holland, but this must be an error.

ECOLOGY
Outside boreal regions the larvae are typical inhabitants of fast-flowing water (Fittkau, 1962).

Procladius   Skuse, 1889

The European species of the genus *Procladius* belong to two subgenera: *Holotanypus* Roback, 1982 and *Psilotanypus* Kieffer, 1906. In the older literature the subgenus *Holotanypus* was regarded as a genus under the name *Procladius*. Because of the many systematic problems within the subgenus *Holotanypus*, the species of this subgenus will be treated together.

Procladius (Holotanypus)   Roback, 1982

Procladius: Goetghebuer, M., 1936: 8–17; Lenz, F., 1936: 62, 73; Fittkau, 1962: 13 vv.; Pankratova, 1977: 83–88; Moller Pillot, 1984: 54.
Procladius (Holotanypus) choreus: Langton, 1991: 12
Procladius crassinervis: Langton, 1991: 10
Procladius culiciformis: Vodopich and Cowell, 1984: 39–52
Procladius ferrugineus: Pankratova, 1977: 86
Procladius (Holotanypus) sagittalis: Langton, 1991: 14
Procladius (Holotanypus) signatus: Langton, 1991: 14

SYSTEMATICS AND IDENTIFICATION
In the classification of Fittkau (1962) the genus *Procladius* belongs to the tribe Macropelopiini. Currently most authors regard *Procladius* and *Tanypus* together as the tribe Procladiini. Other authors classify these as two tribes: Procladiini and Tanypodini. Like most older authors, Fittkau regarded the subgenus *Psilotanypus* as a separate genus. The name *Holotanypus* has been given to his genus *Procladius* (at present a subgenus) because the European species do not belong to the true subgenus *Procladius*.

It appears to be extremely difficult to distinguish between the adult males of all species within the subgenus. Kieffer described tens of species which could not be identified later (see Goetghebuer, 1936: 14–17). Several specialists have tried to put the subgenus in order, but without success. Brundin (1949) made a new key for the Swedish species, again describing several new species. Lehmann (1971) was not able to identify any of his species from the river Fulda, and Pinder (1978) could not distinguish *P. choreus* from *P. culiciformis*. Kobayashi (1998, 2000) stated that the heel height of the gonostyle, regarded as specific, was shorter in summer than in spring and autumn. He supposed that *P. sagittalis* could not be maintained as a species. In any case, the method of identification used by Pinder (1978) does not work. Between the tergite armament of *P. choreus* and *P. sagittalis* given by Langton (1991) a gradual transition exists.
Goddeeris (1983: 137) was also unable to identify his species according to the existing classification. Aagaard (1974) calls attention to the fact that parasitism can cause aberrations in the gonostyle. In the Russian literature *P. ferrugineus* is often reported, but due to a lack of reliable characters this name is never used in Western Europe. The Dutch larval material still gives the impression that two or more species may now fall under *P. choreus*.

In the text on this subgenus we use species names when these are used by the primary authors. Otherwise we use the name *P. choreus* if pupae or exuviae have been identified using Langton (1991). Further, we often refer to *Procladius* larvae (without species name) because in most cases the subgenus or species of the larvae has not has been determined

DISTRIBUTION IN EUROPE AND THE NETHERLANDS
Larvae of the subgenus *Holotanypus* are found throughout the whole of Europe and the

Netherlands. In most Dutch water bodies all or most *Procladius* larvae belong to this sub-genus. Because many water types seem to present a suitable habitat, *Procladius choreus* in particular is among the most recorded Chironomidae in our country. Nijboer and Verdonschot (2001) mention *Procladius* from all 25 regions and from 58.6 % of all localities investigated by our water authorities. Specific information about the occurrence in the Netherlands is given under Water type.

LIFE CYCLE

Belyavskaya and Konstantinov (1956) reported that the numbers of *Procladius* larvae repeatedly display peaks which apparently do not reflect new generations, but differences in food availability. Goddeeris (1983) was unable to unravelling the life cycle of *Procladius* completely, probably due mainly to the problems of identifying populations consisting of more than one species. Nevertheless, he claimed that the larvae go into diapause in winter in the fourth instar. In the Netherlands during winter *Procladius choreus* larvae are mainly in fourth instar, with a smaller percentage in third instar until at least February.

Larsen (1992) stated that *Procladius choreus* was univoltine in two Danish lakes and only flew from late August until the end of November. *Procladius signatus* was also univoltine, but flew during the whole summer. Sokolova (1971) mentioned that *Procladius* in the Ucha reservoir had one generation in the littoral zone, but two in the profundal zone; Laville and Giani (1974) supposed the same was true for Lake Port-Bielh in the Pyrenees. Mundie (1957) found two clear peaks in the emergence of *P. choreus* and *P. crassinervis*, in June/July and in August/September, which were nearly the same in both shallow and deeper water. Pinder (1983) found three generations of *Procladius choreus*, emerging from July to November (in autumn the larvae disappeared from the stream by drift), but Pinder's method may not have been reliable enough for separating the generations. It seems plausible that this species has more than one generation a year because in the Netherlands pupae and exuviae of *P. choreus* have been found from March until October. However, the problems with distinguishing between species mean that this cannot be claimed with certainty. Exuviae of *P. sagittalis* have been found from May until September.

Exceptional life cycles were found in the aerated drinking-water basins in the Biesbosch (Netherlands). Exuviae of *P. crassinervis*, *Procladius* Pe 4, *P. sagittalis* and *P. signatus* have been found here not only in spring and summer, but also in December. Kobayashi (1998) found in Japan, that the adults were smaller in summer than in spring. Drawing on the investigations by Hayashi (1990), Kobayashi mentions that food shortage and concurrence may possibly lead to smaller individuals. The large differences in the length of prepupae found in *Procladius* may therefore be due to environmental factors or to specific differences. According to Sokolova (1968) the larvae grow fastest in the fourth instar in June if enough food is available. She found a mean increase of dry weight per larva of 0.09–0.21 mg in three and a half weeks.

FEEDING

There is a large literature on feeding, especially from Russian and Polish authors (Belyavskaya and Konstantinov, 1956; Tarwid, 1969; Kajak and Dusoge, 1970; Dusoge, 1980; Izvekova, 1980). It appears that the first instar larvae feed on algae, and in later instars mainly on small Crustacea, Chironomidae and Oligochaeta. According to Izvekova (1980) *P. ferrugineus* larvae often switch to feeding on algae also in later instars, in contrast to *P. choreus*. Vodopich and Cowell (1984) observed that third instar larvae of *P. culiciformis* feeding only on algae and detritus were not able to moult to the fourth instar. Baker and McLachlan,(1979) found that fourth instar larvae of *P. choreus* grew successfully on detritus, but were unable to pupate. The larvae actively pursue animal prey, while non-moving food probably is found during creeping with the help of chemoreceptors (see also Chapter 6, General Ecology of the Tanypodinae, p. 72). Small animal prey is swallowed whole; larger prey are sucked, the complete content of the gut being taken in, leaving only the sclerotised parts. Belyavskaya and Konstantinov (1956) observed that the food passes through the gut in 30 minutes to 2 hours. *Procladius* larvae find their prey more easily if it is less able to hide. The biomass of the larvae increases much more in the presence of more prey (Kajak and Dusoge, 1970); *Procladius* can then become more abundant (Kajak et al., 1968).

MICROHABITAT

From the second instar, *Procladius* larvae are pronounced soil dwellers and are found only in small numbers on stones and very rarely on plants. In stagnant water the larvae live in both mineral and organic soils, but in faster running streams almost only on or in soils with coarse and fine detritus (Tolkamp, 1980). However, beside first instar larvae also second and subsequent instars can be present in the water (see under Dispersal). The relatively frequent movements of the larvae of this genus mean that they are sometimes temporarily present outside their optimal microhabitat. Although the larvae rarely live on water plants, they can be found in not too dense vegetation almost as often as on open soil (Steenbergen, 1993). As in many other Tanypodinae, the pupae often shelter in the vegetation at the water's edge.

DENSITIES

Lakes

In many European lakes densities up to 500 larvae/m² have often been found, locally rising to more than 1000 (Brundin, 1949; Kajak et al., 1968; Kajak and Dusoge, 1970; Jónasson, 1972; Bijlmakers, 1983; Lindegaard and Jónsson, 1987; Smit et al., 1996; Lindegaard and Brodersen, 2000). In the drinking-water basins in the Biesbosch, Carpentier et al. (1999) found densities of more than 1000 larvae/m² to be common, with a maximum of 5000/m². These basins have both a very high oxygen content and a rather high productivity (A. Kuijpers, personal communication).

Small rivers

For small rivers in Russia, Balushkina (1987) gives densities from 20 to 330 larvae/m². Similar densities were also found by Moller Pillot (2003) in the Roodloop, a small tributary in the Netherlands.

OVIPOSITION

The eggs are deposited on very different places or thrown off near the banks or above open water. Sokolova (1971) found pelagic egg masses of *P. ferrugineus* over the whole Ucha reservoir. Nolte (1993) recovered an egg mass in the littoral zone of a lake among algal mats.

WATER TYPE

Current

The larvae are often numerous in stagnant and slow flowing waters (current less than 50 cm/s).In faster-flowing water the genus is scarce and it may concern a species characteristic of running water (see Lehmann, 1971). In large rivers in the Netherlands the larvae are dispersed (Klink and Moller Pillot, 1982; Smit, 1995), perhaps because they are carried away by drift (see under Dispersal).

Dimensions

Procladius larvae live in water bodies ranging in size from very small to very large. They can be found in ditches and watercourses about a metre across and pools and cuttings with an area of a few square metres. Verdonschot et al. (1992) state that the larvae are scarce in small stagnant waters. The females may not or rarely fly in woods and therefore rarely deposit eggs in woodland pools. At the other end of the scale the larvae can be found in lakes, rivers and estuaries (see for instance van der Velden et al., 1995). The larvae also live on the bottom of lakes more than 10 m deep, even at a depth of 50 m or more (Brundin, 1949: 518). At great depths oxygen deficit can be a problem, but if not, the highest densities of larvae can be observed at greater depths, as in a lake in northern Norway (Aagaard and Sivertsen, 1980). Vodopich and Cowell (1984) report that in one lake P. culiciformis was most abundant in the littoral zone (< 1 m depth) and very scarce at depths below 3 m, but in another lake the larvae were most abundant at the deep stations. In these cases the key factor was the presence of food.

Permanence
The larvae of Procladius are often found in temporary pools, peat cuttings, ditches and the upper courses of small rivers (e.g. Ketelaars, 1986). Because they cannot survive total desiccation of their environment, Dettinger-Klemm (2003: 74) classifies them as colonisers. Of course, recolonisation is only possible during periods in which the females are flying and not in late autumn and winter (Werkgroep Hydrobiologie MEC Eindhoven, 1993; Moller Pillot, 2003). Shallow trenches, therefore, are perhaps not usually colonised.

Specific differences
The majority of the pupae and exuviae found in the Netherlands (identified using Langton, 1991) belong to P. choreus, especially in running water and in small stagnant water bodies. P. sagittalis is common in different types of lakes (for instance dune lakes). In some cases, P. signatus and Procladius Pe 1 and Pe 4 have been found. P. crassinervis has been quite common in drinking-water basins in the Biesbosch since 1997 (A. Kuijpers, unpublished).

SOIL
Vodopich and Cowell (1984) found that P. culiciformis larvae have a highly significant preference for organic sediment above more mineral sediments. The larvae also grew better on the organic sediment, unless oxygen concentrations were low. According to Steenbergen (1993) Procladius is a little scarcer on clay than on sand and peat, but we found locally the larvae to be very numerous on clay.

pH
In many types of water, the occurrence of Procladius larvae seems to be independent of pH. Leuven et al. (1987) found them to be fairly numerous at pH < 4.0 and still present at pH 9.45. However, densities are probably lower in acid water because less food will usually be available. In the eutrophic Belversven lake, Bijlmakers (1983) incidentally found densities of 600 larvae/m² (for instance on the bottom between reed), but in the very acid Staalbergven lake (pH between 4 and 5) the maximum density was 325 larvae/m².

TROPHIC CONDITIONS AND SAPROBITY
In the province of North Holland the larvae of Procladius occur abundantly in phosphate-poor and nitrogen-poor as well as hypertrophic water, although they occur a little more frequently in phosphate-poor water (Steenbergen, 1993). In lowland brooks the larvae are less common after heavy pollution, although they can survive for a long time in water almost devoid of oxygen after transport by drift (Moller Pillot, 1971). Moller Pillot and Buskens (1990) and Verdonschot et al. (1992) do not give clear differences in occurrence between water with different trophic conditions. On the other hand, the occurrence and numbers of Procladius larvae are largely determined by the availability of food (Jónasson, 1977; Vodopich and Cowell, 1984) and therefore indirectly dependent on total production. Accessibility of the prey is also a decisive factor (Kajak et al., 1968; Kajak and Dusoge, 1970).

Vodopich and Cowell (1984) report that the larvae of *P. culiciformis* grow very well at a dissolved oxygen level of about 2 mg/l. Jónasson (1972: 88) mentions that the larvae are badly adapted to low oxygen concentrations, although he found that they survived more than a month experimentally in oxygen-free water, and the larvae emigrated from the profundal zone of Lake Esrom during summer and survived in shallower areas. We also found few *Procladius* larvae on the bottom of Dutch ditches with long nocturnal anoxia. It is very striking that in the basins in the Biesbosch *P. choreus* is very rare but several other species of the genus (also subgenus *Psilotanypus*) are common. Compared with most other water bodies in the Netherlands the oxygen content in these basins is very high, but nevertheless production is not low (the basins are aerated). This may be the reason for the presence of *P. crassinervis*, which is not known to be present in other Dutch water bodies. Orendt (1993) found *P. crassinervis* in Bavaria only in oligotrophic environments.

CONTAMINANTS

Warwick (1989, 1991) states that morphological deformities in *Procladius* larvae occurred frequently in waters known to be seriously contaminated by PCBs and heavy metals. However, *Procladius* seems to be more tolerant than *Chironomus*. Deformities in *Procladius* are most striking in ligulae and paraligulae, but occur also in antennae, mandibles and mentum. For the identification of the larvae this is no problem. The deformities are nearly always asymmetrical.

SALINITY

The larvae of *Procladius choreus* endure light brackish (oligohaline) water very well. In Zeeland they are found in water bodies with a chloride content up to 4400 mg/l (Parma and Krebs, 1977; Krebs and Moller Pillot, in preparation). In North Holland the larvae were much scarcer in water with chloride contents above 1000 mg/l (Steenbergen, 1993). In Lake Volkerak-Zoommeer they occurred in large numbers only after freshening (van der Velden et al., 1995). In the Camargue, however, they endure a much higher content (up to 9300 mg/l: Tourenq, 1975) and in Estonia they also live in mesohaline water at concentrations up to 6800 mg/l (Tõlp, 1971). The difference in tolerance found in the Netherlands and elsewhere can be explained by the great fluctuations in salinity in the Netherlands. Parma and Krebs (1977) also reported the presence of *P. sagittalis* (as *P. breviatus*) in brackish water.

DISPERSAL
Larvae

As in other genera, the larvulae lead a planctonic existence. However, Koreneva (1959) observed that they are more frequent near the bottom than other Tanypodinae larvulae. Mundie (1965) and Sokolova (1968) caught more older larvae in the water than larvae of other species (up to 10 m above the bottom), especially in April and June. The reason for this behaviour was not water movement, but possibly oxygen depletion at the bottom. Romaniszyn (1950) observed an annual migration of *Chironomus* and *Procladius* larvae to the profundal zone of a lake in autumn and a return to the littoral zone in early spring (see also Thienemann, 1954: 701–702). The reason for this migration could be to seek water with an optimal temperature and oxygen content. In Dutch lowland brooks the larvae are probably transported over large distances by drift (Moller Pillot, 1971; 2003). In the Roodloop this was found especially in winter when the current was strong and food and shelter were less available. Klink (1990) remarks that *Procladius* larvae are easily transported in brooks and rivers, especially after a change in water level.

Adults

The rapid colonisation of isolated water bodies is without doubt due to the common occurrence of the genus as well as the easy dispersal of adult females by air. Accordingly, Moller

Pillot (2003) attributes *Procladius* to the circulating taxa, which remain present in the landscape as a whole because of their propensity to move in dynamic environments.

Procladius (Psilotanypus) flavifrons   (Edwards, 1929)

The males and pupae of this small species are easy to identify (Goetghebuer, 1936; Pinder, 1978; Langton, 1991). The larvae are very small and may incorrectly be identified as third instar larvae. The species is known from European lakes. Brundin (1949) found *P. flavifrons* especially in oligotrophic lakes in Sweden. Two finds have been recorded in the Netherlands: a moorland pool in the province of Noord-Brabant (H. van Kleef) and the drinking-water basins in the Biesbosch (A. Kuijpers, personal communication). The adults emerged in May, July and August.

Procladius (Psilotanypus) lugens   (Kieffer, 1915)
Procladius (Psilotanypus) imicola   (Kieffer, 1922)

IDENTIFICATION
The species *P. lugens* and *P. imicola* are treated together here because there is some doubt about whether these two species are always properly and consistently identified in the literature. According to Goetghebuer (1936) adult males can be identified easily by the colour of the cross vein Mcu and their antennal ratio. *P. imicola* is absent from Pinder (1978) and Langton (1991) and the description of the pupa in Pankratova (1977) is insufficient for identification. The larva of *P. lugens* has not been described. Differences between the larvae of *P. imicola* and *P. rufovittatus* are described in Pankratova (1977). The claws of the posterior parapods of *P. imicola* do not have teeth, as Fittkau and Roback (1983: 64) suggest.

DISTRIBUTION IN EUROPE AND THE NETHERLANDS
P. lugens has been recorded in the British isles, Finland and Germany (Fittkau and Reiss, 1978; Ashe and Cranston, 1990; Chandler, 1998; Samietz, 1999). *P. imicola* is only known from Germany, Sweden, Poland and Russia (Kajak et al., 1968; Fittkau and Reiss, 1978; Ashe and Cranston, 1990; Samietz, 1999). Males of *P. lugens* identified using Pinder (1978) have been caught in the Netherlands near the Maarsseveen lake (Kouwets and Davids, 1984) and in the Groote Peel (unpublished). Exuviae (identified using Langton, 1991) have been found in the Afgedamde Maas near Andel (unpublished).

LIFE CYCLE
In the reservoirs studied by Mundie (1957) *P. lugens* was clearly univoltine. Almost all adults emerged in May. Sokolova (1971) also calls *P. imicola* univoltine and found prepupae (near Moscow) in April. Koreneva (1959) mentions the emergence of this species in the first half of June.

FEEDING
According to investigations by Izvekova (1980) the larvae feed mainly on animals. Copepoda play an important role in their diet alongside Cladocera, Ostracoda, Rotifera and Chironomidae. Algae (mainly diatoms?) and detritus are also eaten. The larvae cannot complete their development without animal food.

OVIPOSITION
Sokolova (1971) observed the egg masses of *P. imicola* in the open water of lakes. They are

thrown off above the open water surface far from the water's edge, where the water is deep (Koreneva, 1959: 118). Koreneva counted 200–270 eggs per egg mass.

MICROHABITAT

The larvae of *P. imicola* live in the uppermost silt layer, where they creep around in search of prey (Izvekova, 1980).

DENSITIES

Kajak et al. (1968) found more than 1000 *P. imicola* larvae/m² in Polish lakes. However, such densities seem to be more exceptional. Balushkina (1987) found about 200 larvae/m² in lakes near Leningrad in spring and about 1000 young larvae/m² for a short time in July.

WATER TYPE

P. lugens and *P. imicola* are found mainly in lakes. According to Mundie (1957) the larvae live in shallow as well as deeper areas, the optimum depth being approximately 5 m. A Dutch record of *P. lugens* in the Groote Peel shows that this species can also live in small, shallow acid water bodies. Shilova (1976) found only a few individuals of *P. imicola* in old river channels around the Rybinsk reservoir. The species was not encountered in the reservoir itself. *P. imicola* has been found in the profundal zone of the Uchinsk reservoir near Moscow (Koreneva, 1959; Sokolova, 1971).

Procladius (Psilotanypus) rufovittatus (Van der Wulp, 1873)

IDENTIFICATION

Adult males of *P. rufovittatus* are easily recognised by the shape of the gonostyle. They differ in this character also from *P. imicola*, which is not treated by Pinder (1978). Identification of pupae and exuviae is possible using Langton (1991), but also here *P. imicola* is lacking. If *P. rufovittatus* can be recognised by the shape of the anal lobe, as written by Pankratova (1977), deserves further investigation. Pankratova (1977: 88, fig. 31) characterises the larvae by a short claw of the posterior parapods with teeth. Larva and pupa are more extensively described by Muragina-Koreneva (1957), who gives a head width of 0.49–0.52 mm. The species is probably smaller than most other species of the genus.

DISTRIBUTION IN EUROPE AND THE NETHERLANDS

P. rufovittatus is known from Western and Central Europe, Finland and Russia (Ashe and Cranston, 1990). In the Netherlands there are few data, mainly because the larvae of the genus *Procladius* could not be identified. Several records of adult males and exuviae suggest that the species occurs dispersed over the country.

LIFE CYCLE

In Russia Muragina-Koreneva (1957) found only one generation a year, emerging in July. Shilova (1976) claims this must be an error, having found prepupae in the Rybinsk reservoir mainly in June and the end of August and concluded that there are two generations a year. In England Mundie (1957) found only one generation, emerging from early June to August, with one clear peak in June in shallow water and in July in deeper water. He thinks that a bimodal distribution can be caused by environmental factors or a larval diapause. A similar bimodal curve was found by Schmale (1999) in the dune lakes in Berkheide (the Netherlands). Orendt (1993) mentions two or three generations for German lakes, whereas Lindegaard and Brodersen (2000) found one clear peak in a Danish lake in 1993 at the end of May and in 1994 at the end of June. In the Biesbosch (Netherlands) exuviae were sampled every month from the end of April until the end of August (Kuijpers, unpublished). Janecek

(1995) reported one generation a year in a carp pond in Austria. The overwintering larvae were mainly third instar, but smaller numbers were second and fourth instar.

FEEDING
There are no data available on feeding, but the diet of this species is probably very similar to that of *P. imicola* (see above).

DENSITY
Lindegaard and Brodersen (2000) found 140 larvae/m² in a Danish lake. However, the densities found in Kempton Park East Reservoir in England were considerably higher (Mundie, 1957). Janecek (1995) counted nearly 12,000 young larvae/m² in late summer and 500–2000 full-grown larvae in spring.

WATER TYPE
All literature data refer to stagnant, not very small and usually deep water bodies (cf. Fittkau and Reiss, 1978). The Dutch records are consistent with this (Lake Maarsseveen, Biesbosch, dune lakes). Mundie (1957) found the larvae at depths of between 1 and 8 m in nearly the same densities. However, the species has also been found in the Guitjesdel dune lake, which is no deeper than 1 m (Schmale, 1999) and in a moorland pool near Soerendonk (H. van Kleef). In the Rybinsk reservoir Shilova (1976) found larvae living at a range of depths from the littoral zone down to a depth of 18 m).

MICROHABITAT
According to Muragina-Koreneva (1957) the larvae live on the water bottom, the highest densities being at a depth of 5–10 m. Shilova (1976) found the larvae mainly on silty soil and on sandy soil with silt.

SWARMING AND OVIPOSITION
Muragina-Koreneva (1957) observed that the males formed dense globular swarms at a height of 3–4 m. The females usually rest in bottom vegetation or shrubs. The egg masses are thrown off on the whole water surface, but mainly near the shore. They contain 150–250 eggs (Koreneva, 1959).

TROPHIC CONDITIONS
In the Netherlands the species has been found in more or less mesotrophic conditions. Orendt (1993) found the larvae most abundant in eutrophic lakes in Bavaria.

Psectrotanypus varius   (Fabricius, 1787)

SYSTEMATICS
Within the tribe Macropelopiini the genus *Psectrotanypus* has some plesiomorphic characteristics, such as the many small teeth on the mandible (Fittkau, 1962). *P. varius* is the only species in the genus found in Europe.

DISTRIBUTION IN EUROPE AND THE NETHERLANDS
P. varius lives throughout Europe (Fittkau and Reiss, 1978). In the Netherlands it is a common inhabitant of stagnant and slowly flowing water (Moller Pillot and Buskens, 1990; Verdonschot et al., 1992).

LIFE CYCLE
Pupae of *P. varius* have been found in the Netherlands from March to October. It must be

assumed that they have a diapause in autumn because hardly any fourth instar larvae are found in the second half of October and in November. However, in December many of the larvae are fourth instar. Second instar larvae are still found in January and third instar larvae are found until March. As a result of these differences in development, emergence is not synchronised; the first generation emerges from the second half of March until the middle of May. Fourth instar larvae may be absent for a short period, but only locally. Between March and October probably three or more generations emerge, at least in small stagnant water bodies.

FEEDING

According to Belyavskaya (1956) young larvae of *P. varius* feed primarily on diatoms and green algae; older larvae feed almost only on animal food. Small prey is swallowed whole and larger prey is sucked out. Chironomidae are probably the most important source of food. Krenke (1968) observed that older larvae also often eat more algae than animal food. The larvae probably sometimes eat detritus rich in bacteria, as has been observed in other Tanypodinae (see p. 72).

MICROHABITAT

The larvae live on the bottom of stagnant and slowly flowing waters, often on and between silt and dead organic material. The greatest densities have been found by us between grassy vegetation where silt accumulates. On plants, stones and wood the larvae can be found only near the bottom.

DENSITIES

Rather large numbers of larvae can accumulate at optimal sites, for instance where silt accumulates between grasses in small lowland streams or in small pools reduced by evaporation. In summer and autumn from 400 to more than 1000 larvae/m^2 can be present temporarily (Beattie, 1978; Balushkina, 1987; Moller Pillot, 2003), but high densities of larvae are found only locally. In flowing water densities decrease as discharge rates increase in autumn and winter (Moller Pillot, 1971; 2003).

WATER TYPE

Current

The species is a characteristic inhabitant of stagnant water (Fittkau, 1962). In running water occurrence is mostly restricted to places or periods with slower currents (Lehmann, 1971; Moller Pillot, 1971, 2003).The larvae live especially where organic silt is deposited (see under Microhabitat).Sometimes the species has been found in springs (Lindegaard, 1995; Verdonschot, 2000: 60).

Dimensions

The larvae are found most frequently in water bodies less than 4 m wide and 30 cm deep, with numbers decreasing as width and depth increase (Steenbergen, 1993). The larvae are also found in very small rain puddles.

Permanence

The larvae have no protection against total desiccation of the soil, but often occur in temporary water. The common occurrence of this species and the long flying period encourage rapid colonisation after a wet period, aided by the fact that ovipositing females probably select small water bodies (Dettinger-Klemm, 2003: 214).

pH

The species is usually absent from very acid environments and organic soil (Verdonschot et al., 1992; Werkgroep Hydrobiologie, 1993; Moller Pillot, 2003). *P. varius* may occasionally be numerous at pH < 5.0 (see for instance Schleuter, 1985: 25/91). In these places there is generally much decomposition and the pH within the soil is probably much higher and/or the soil is rich in nitrogen.

TROPHIC CONDITIONS AND SAPROBITY; OXYGEN CONTENT

In lowland brooks *P. varius* is a characteristic inhabitant of organically polluted water (Fittkau, 1962; Moller Pillot, 1971; Peters et al., 1988). The larvae survive very low oxygen contents for a long time, and can even tolerate anoxia even for some time (Moller Pillot and Buskens, 1990). In stagnant water the optimum condition is less sharply defined, although Steenbergen (1993) mentions a pronounced increase in occurrence in water with a higher ammonium content. The main point seems to be that the larvae benefit from decomposition of organic material, for older larvae especially because of the presence of more prey. The species is not mentioned in many lists of saprobity indicators, probably because these lists are made for relatively fast-flowing water or because chironomids were not identified.

SALINITY

Parma and Krebs (1977) did not find *P. varius* in Zeeland at a chloride content > 1000 mg/l. However, in the province of North Holland the larvae have been found in slightly more saline water, but infrequently (Steenbergen, 1993). Verdonschot et al. (1992) mention the species also for polyhaline water, but this must be an error. At chloride contents > 3000 mg/l the larvae are nearly absent (Moller Pillot and Buskens, 1990).

TOXIC CHEMICALS

The larvae are very tolerant of iron: Rasmussen and Lindegaard (1988) found them in water with ferrous iron concentrations of more than 4 mg Fe^{+2}/l.

DISPERSAL

In landscapes where *P. varius* is common, immigration by air appears to occur often. In brooks many larvae are transported by drift, especially when appropriate microhabitats and food are in short supply locally, for instance in winter (Moller Pillot, 2003). Larvae and pupae can be transported by drift over some kilometres (Moller Pillot, 1971: 199).

Rheopelopia   Fittkau, 1962

SYSTEMATICS AND RELATIONSHIP

According to Fittkau (1962) the genus *Rheopelopia* is closely related to *Conchapelopia* and these two genera belong with *Thienemannimyia* and *Arctopelopia* to a group of genera defined by him as the *Thienemannimyia*-Reihe. Owing to the strong resemblance of the larvae Moller Pillot (1984) grouped these genera together as 'cf. *Conchapelopia*'. In this book we call them *Conchapelopia* aggregate.

THE EUROPEAN SPECIES

Three species of the genus occur in Europe: *R. ornata* (Meigen, 1838), *R. maculipennis* (Zetterstedt, 1848) and *R. eximia* (Edwards, 1929). In the Netherlands only *R. ornata* has been found, but *R. maculipennis* may well be present as well (see under Distribution).

IDENTIFICATION

A pupa of *R. ornata* found in southern Limburg had a thoracic horn about ten times longer than it was broad, contrary to the description by Langton (1991: 30).

DISTRIBUTION IN EUROPE AND THE NETHERLANDS

The whole genus is confined to running water. *R. ornata* is known from nearly the whole of Europe. In the Netherlands this species lives nearly exclusively in larger rivers and rarely in small rivers. *R. maculipennis* occurs in Northern and Central Europe and in England (Fittkau and Reiss, 1978) and also lives in rivers, probably not in the lower reaches. This species occurs in the Rhine only in southern Germany (Caspers, 1991). *R. eximia* is a rare species, only known from Great Britain, Ireland and the Black Forest (Fittkau, 1962).

LIFE CYCLE

In the Netherlands Klink (1982) found adults of *R. ornata* from May to August. In the river Worm a pupa was sampled as late as September (B. van Maanen, unpublished). Records from other authors (Fittkau, 1962; Lehmann, 1971) fall within this period. It is not clear if there are one or two generations a year.

FEEDING AND BEHAVIOUR

Klink (1982) found no chironomids in the guts, only chaetae of Naididae and Tubificidae, which are probably the most important prey. The larvae creep around in the silty layer on plants and stones in the river, between sponges, Bryozoa and tubes of chironomids such as *Rheotanytarsus* or *Dicrotendipes nervosus*.

MICROHABITAT

The larvae live mainly on the upper surfaces of stones and can be numerous on artificial substrates. They can be found at lower densities on silted plant stems and on sandy bottoms in shallow water (Lehmann, 1971; Smit, 1982; Smit and Gardeniers, 1986; Klink, 1991). Klink (1982) found the pupae in the tubes of other chironomids such as *Dicrotendipes nervosus*. He points out that the pupae of other Tanypodinae move freely in the water column and that the typical form of the thoracic horn of *Rheopelopia* may be an adaptation to living in strong currents.

WATER TYPE

Current and dimensions

Fittkau (1962) calls *R. ornata* a typical inhabitant of the fast running middle reach of rivers (*Barbenregion*). According to Caspers (1980) *R. ornata* is the most numerous of the Tanypodinae in the River Rhine near Bonn. In the Netherlands the species has been found mainly in the large rivers (Rhine, Waal, IJssel and Meuse) and rarely in small rivers. Smit and Gardeniers (1986) found the larvae in the Meuse only where the current was faster than 0.4 m/s.

Schineriella schineri   (Strobl, 1880)

Pentaneurini gen. *schineri* Fittkau, 1962: 275
Krenopelopia schineri Fittkau and Reiss, 1978: 407; Pinder, 1978: 32, fig.83A.
Tanypodinae genus II Fittkau and Murray, 1986: 65, fig. 5.47.

SYSTEMATICS RELATIONSHIP

For a long time the species was thought to be related to *Krenopelopia*, but the genus seems to be more related to *Labrundinia* (Murray and Fittkau, 1989).

IDENTIFICATION OF ADULTS

The adult male is almost impossible to identify using Pinder (1978). The easiest character to look for is the absence of the R_{2+3} in *Schineriella*. The posterior abdominal tergites may exhibit brownish bands (see Murray and Fittkau, 1989).

DISTRIBUTION IN EUROPE AND THE NETHERLANDS

According to Fittkau and Reiss (1978) *S. schineri* is known from large parts of North-Western and Central Europe. The species has been found in different parts of the Netherlands. To date it is known from few localities because the larva has not been described. In the past most larvae have probably been identified as *Zavrelimyia*.

LIFE CYCLE

Pupae are found in the Netherlands in May. In October larvae were in the third and fourth instar. Goetghebuer (1911: 106) caught adults in June and August (as *Ablabesmyia Schineri*).

WATER TYPE

According to Reiss (1984) *S. schineri* lives mainly in ponds and pools. In Britain (Wilson and Ruse, 2005) and the Netherlands the larvae have been found in ponds and lakes, rarely in very slowly flowing water.

TROPHIC CONDITIONS AND SAPROBITY

According to Wilson and Ruse (2005: 72, 157) the larvae live in nutrient-rich ponds and lakes, but they are intolerant of organic pollution. In the Netherlands larvae have been found more or less regularly in mesotrophic lakes (Botshol, Naardermeer) and also in eutrophic water bodies elsewhere in the country.

Tanypus kraatzi   (Kieffer, 1913)

Pelopia kraatzi: Chernovskii, 1949: 156; translation 1961: 262.
Pelopia punctipennis: Koreneva, 1959, 1960 (nec aliis)
? *Tanypus punctipennis*: Beattie, 1978: 78 etc. (nec aliis)

IDENTIFICATION OF ADULTS

Adult males cannot be identified using Pinder (1978) because this species is not included in this key. See also under *T. punctipennis*.

DISTRIBUTION IN EUROPE AND THE NETHERLANDS

According to Fittkau and Reiss (1978) *T. kraatzi* is absent from the British Isles, Scandinavia and large parts of Southern Europe. However, this is partly due to incorrect identification of the adults (see above). In the Netherlands the species can be found dispersed over the whole country (Nijboer and Verdonschot, 2001), but is obviously much more common in the Holocene areas in the west and north.

LIFE CYCLE

In the Netherlands pupae and exuviae have been found from May until October. It seems likely that there are two generations a year in the Netherlands, although Koreneva (1960, as *Pelopia punctipennis*) reported only one generation near Moscow. Most larvae in the Netherlands are in the second (and third) instar during the whole winter (up to early April), with only a few in the fourth instar (own observations). Clearly there is a winter diapause. Beattie (1978: 78) also states that the larvae in the Tjeukemeer lake (as *T. punctipennis*, however probably *T. kraatzi*) hibernate (indiapause) in the second instar.

FEEDING

Considering the very similar structure of the mouth parts, the larvae of the various species of the genus *Tanypus* will probably exhibit no or hardly any differences in their food preference. The genus will be treated here as a whole. In the gut of *Tanypus* larvae Roback (1969) and Titmus and Badcock (1981) found almost exclusively algae (mostly unicellular) and plant remains. Tarwid (1969) also states that the gut contents consist almost entirely of unicellular algae, with rarely Cladocera or their eggs. Biever (1971) states that *Tanypus grodhausi* is definitely not a predator. In the gut of *T. kraatzi* sampled in the Netherlands we found diatoms, green and blue-green algae and detritus rich in bacteria. Titmus and Badcock (1981) observed that the larvae only lived where the relevant algae grew and for this reason concentrated on the surface of the sediment. According to Konstantinov (1958), however, older *Tanypus* larvae can eat relatively many other Chironomidae. Adding larvae of *T. punctipennis* to an aquatic system in the laboratory led to a sharp decrease in the numbers of Copepoda and Ostracoda (Kajak et al., 1968). Moog (1995) writes that *Tanypus* larvae live mainly on animal prey and also detritus, but this estimation is probably based on too little information. The larvae appear to feed mainly on algae and detritus, but they use animal food (small Crustacea or Chironomidae) when it is readily available.

MICROHABITAT

The larvae of *T. kraatzi* are strongly pelophilous. In the Netherlands and elsewhere they are especially found on silty bottoms, often even on thick layers of organic silt. They concentrate on the surface of the sediment because this is where the greatest numbers of unicellular algae can be found (Titmus and Badcock, 1981). Pupae, however, are mostly sampled between vegetation and at the surface of the water.

DENSITIES

In lakes and ditches the larvae are often found in densities of hundreds per square metre, sometimes up to (locally?) 3300 young larvae/m², but are less numerous after overwintering (Kajak et al., 1968; Beattie, 1978; Bijlmakers, 1983). Some of these densities possibly also apply to *T. punctipennis* (identified using Pinder, 1978).

OVIPOSITION

According to Koreneva (1959, 1960, as *Pelopia punctipennis*) egg-shaped egg masses were thrown off in the open water of the studied reservoir. She found approximately 600 eggs per egg mass.

WATER TYPE

Current

Although the species is not adapted to living in running water, larvae have been found incidentally in such water, especially in slowly flowing canals (Steenbergen, 1993). *T. kraatzi* is absent in the rivers Rhine (Caspers, 1980) and Meuse (Smit, 1982; Klink, 1991).

Dimensions

The larvae can be abundant in narrow ditches (2 to 3 m wide) and in large lakes. *T. kraatzi* is clearly less stenotopic as other species of the genus. In the province of North Holland *T. kraatzi* is most common in water bodies less than 10 m wide, although the larvae are not scarce in larger water bodies (Steenbergen, 1993). Koreneva (1960) found the larvae (as *P. punctipennis*) at a depth of 0.5–3 m, but not in the deeper parts of the Ucha reservoir.

Permanence

In temporary water larvae can be found after oviposition or flooding. The larvae are not able to withstand complete desiccation.

SOIL

According to Steenbergen (1993) in the province of North Holland *T. kraatzi* is a little more common on peat than on sand and clay. In other parts of the country the species has been found relatively often in ditches on peaty subsoil (van Gijsen and Claassen, 1978; Hermans and Kahmann, 1985).

pH

Roback (1974) mentions the presence of *Tanypus* larvae only at fairly high pHs (7.4–8.4). In the Netherlands *T. kraatzi* is also usually found at a pHs between 7.2 and 8.3. Few catches are known from water with a pH between 6 and 7, and hardly any from more acid water (e.g. Verstegen, 1985).

TROPHIC CONDITIONS AND SAPROBITY

Because the larvae require a layer of organic silt and a fairly high pH, an important condition seems to be that much food is available. Nevertheless, Verdonschot et al. (1992) consider *T. kraatzi* to be a fairly characteristic species of mesotrophic water. However, Steenbergen (1993) found the larvae most often in water with the highest concentrations of phosphate and ammonium (orthophosphate > 1.5 mg P/l; ammonium > 1.6 mg N/l) and the lowest oxygen contents. Furthermore, the larvae were also common in water with less than 0.05 mg orthophosphate-P/l and almost no dissolved nitrate and ammonium. Our own unpublished data suggest that the larvae survive well if for some hours in the night the oxygen saturation percentage is lower than 5%. The genus *Tanypus* is absent or very scarce when little organic material is present, for example in most deep sand pits and storage reservoirs (Mundie, 1957; Buskens and Verwijmeren, 1989; Ketelaars et al., 1993).

SALINITY

According to Steenbergen (1993) the larvae of *T. kraatzi* occur frequently in water bodies with a chloride content between 300 and 1000 mg/l and occasionally also at a chloride content between 1000 and 3000 mg/l. They seem to be absent in water with higher salinity (Moller Pillot and Buskens, 1990).

Tanypus punctipennis   Meigen, 1818

Pelopia punctipennis: Chernovskii, 1949: 155–156, fig. 142; translation 1961: 262.

IDENTIFICATION OF ADULTS

Adult males identified using Pinder (1978) may be ascribed to *T. kraatzi* because this species is not included in this key. The veins of the wing are the most reliable characteristic (Fittkau, personal communication). Also, the key by Goetghebuer (1936: 7) leads to mistakes. It is accepted that both species have been mistaken throughout Europe.

DISTRIBUTION IN EUROPE AND THE NETHERLANDS

T. punctipennis has been found throughout almost the whole of Europe (Fittkau and Reiss, 1978). In the Netherlands the species is common in the Holocene regions in the west and north. Only few records are known from the Pleistocene regions.

LIFE CYCLE

According to Shilova (1976) in the region north of Moscow *T. punctipennis* has two generations a year, flying in June and August. She supposes that Koreneva (1960) incorrectly concludes that the species has one generation a year. In fact, the population studied by

Koreneva belonged to *T. kraatzi* as appears from her description of the larva. In the Netherlands pupae have been found in summer and late summer. The annual cycle most probably corresponds with that of *T. kraatzi* (see there).

MICROHABITAT
As stated for *T. kraatzi*, the larvae of *T. punctipennis* are pronounced pelophilous inhabitants of the water bottom. Shilova (1976) found the larvae at shallow depths.

FEEDING
See under *T. kraatzi*.

DENSITIES
See under *T. kraatzi*.

SWARMING AND OVIPOSITION
In August Shilova (1976) observed a fairly dense swarm of enormous dimensions above a lake bank where the larvae lived. The swarm was more than 40 m long, 2 m high and several metres wide. Egg masses were sampled by Nolte (1993) on silty soil at the bottom of the littoral zone of a lake. Each globular egg mass contained between 560 and 1100 eggs.

WATER TYPE
Current
In the Netherlands the species seems to be confined to stagnant water, but in Germany the species has been found in the lower stretches of rivers and even in fairly strong currents (Reiss, 1968; Lehmann, 1971; Braukmann, 1984; Caspers, 1991; Orendt, 2002). At least in some of these cases there can be no confusion with *T. kraatzi*. Langton (1991) even mentions that *T. punctipennis* is commonest where there is a slow current.

Dimensions
Most recorded samples in the Netherlands are from broad ditches and lakes. Steenbergen (1993) found *T. punctipennis* less often in larger lakes than *T. kraatzi* and less often in narrow ditches (width < 4 m).

SOIL
In contrast to *T. kraatzi* the larva of *T. punctipennis* are found especially on clay (van Gijsen and Claassen, 1978; Steenbergen, 1993).

pH
T. punctipennis is not known from acid water. The species is scarce below pH 7.0 (see e.g. Steenbergen, 1993).

TROPHIC CONDITIONS AND SAPROBITY
Saether (1979) states that *T. punctipennis* is common in the littoral zone of eutrophic lakes and scarce in mesotrophic lakes. In the Netherlands *T. punctipennis* is characteristic of eutrophic, hypertrophic and brackish water, but is not totally absent from mesotrophic ditches and lakes (Verdonschot et al., 1992; Steenbergen, 1993). The latter author mentions that the species is less common in oxygen poor water than *T. kraatzi* , but this can be related to its frequent presence in shallow ditches with no duckweed cover.

SALINITY
T. punctipennis occurs in the Netherlands relatively often in brackish water and can endure a higher salinity than *T. kraatzi*. The larvae are found relatively often in water bodies with a

chloride content above 1000 mg/l (Steenbergen, 1993). Other data from the literature have to be used carefully because adults have often been identified using Pinder (1978).

Tanypus vilipennis   (Kieffer, 1918)

Pelopia villipennis: Chernovskii, 1949: 154–155, fig.141; translation 1961: 262.

DISTRIBUTION IN EUROPE AND THE NETHERLANDS
According to Fittkau and Reiss (1978) *T. vilipennis* has not been found in Scandinavia and in large parts of southern Europe. Edwards (1929) states that *T. vilipennis* is rare in England. In the Netherlands the species is considerably scarcer than the other species of the genus (Nijboer and Verdonschot, 2001). The species has been recorded especially in the fen-peat region in the west of the Netherlands and in the province of Overijssel, with a few records from the floodplain of the main rivers in the central part of the Netherlands and the regions with sandy soils.

LIFE CYCLE
The life cycle of *T. vilipennis* is probably no different from that of *T. kraatzi*. However, a number of fourth instar larvae have been found in October.

OVIPOSITION
The egg masses are thrown off in open water and contain approximately 600 eggs per egg mass (Koreneva, 1959).

MICROHABITAT
Koreneva (1960) found prepupae in the shallow parts of the Ucha reservoir at a depth of 2–3 m. Zinchenko (1997) reports finding this species on silty clay and sandy bottoms with plant remains at a depth of 0.4–1.8 m.

WATER TYPE
T. vilipennis is usually only mentioned as living in stagnant lakes and ponds (Pankratova, 1977; Fittkau and Reiss, 1978; Verdonschot et al., 1992). However, Zinchenko (1997) found the larvae to be fairly abundant on the bottom of the Russian river Chapaevka. In the Netherlands larvae have been found in ditches and lakes. The scarce data indicate that this species, more than the other species of the genus, is confined to water with fairly good oxygen content. It is noteworthy that in the Netherlands *T. vilipennis* occurs especially in the fen-peat regions, which are called 'seepage regions' by Schroevers (1977).

Telmatopelopia nemorum   (Goetghebuer, 1921)

SYSTEMATICS AND RELATIONSHIP
Notwithstanding the similarity between the larvae of *Xenopelopia* and *Telmatopelopia*, these genera are probably not related (Fittkau, 1962; Saether, 1977).

THE EUROPEAN SPECIES
In Europe only one species has been described (Fittkau, 1962). However in Belarus two different larval types could be discerned (unpublished data).

DISTRIBUTION IN EUROPE AND THE NETHERLANDS
T. nemorum has been found in the greater part of Europe, with the exception of the

Mediterranean area (Fittkau and Reiss, 1978). In the Netherlands the species seems to be confined to the Pleistocene region of the country and some surrounding marshy areas.

LIFE CYCLE

In principle, the species has one generation a year, with a summer diapause (as egg or juvenile larva). The larvae reach the fourth instar from November until the end of March. Pupae and adults can be found in April and May (mainly from end of April). In the Netherlands fully grown larvae or pupae have been found fairly often from July until September. It is not clear whether these are late specimens or a partial second generation. In Northern and Eastern Europe larvae are found in summer more frequently; Brundin (1949: 683) caught adult males in Sweden in May and July.

FEEDING AND BEHAVIOUR

There is no literature about the feeding habits of this species. We have usually found much detritus and often some diatoms and other small particles in the gut of the larvae, but no animal remains. Most probably their feeding preference is similar to that of the highly similar *Xenopelopia* larvae. In fright the larvae recoil quickly, like most other Pentaneurini.

MICROHABITAT

As far as is known the larvae live mainly between and on dead organic material on the bottom, such as dead leaves and grasses.

DENSITIES

Fittkau (1962: 285) mentions the mass occurrence of larvae in deciduous woodland pools. In the Netherlands high densities have never been reported; the number of larvae is usually well below 100/m^2. However, Waajen (1982) found more than 100 (up to 250?) larvae in a net sample of approximately 1 m^2 from some peat cuttings in Liesselse Peel.

WATER TYPE

T. nemorum is a characteristic inhabitant of small, stagnant, mostly acid temporal pools such as peat cuttings, pools and ditches in woodland and bogs (Kreuzer, 1940; Verdonschot et al., 1992). They can be found sometimes in less acid (rarely also permanent) marshes or in larger pools. The larvae can arrive into the upper parts of lowland brooks by drift. In Sweden the larvae also live in lakes (Brundin, 1949).

pH

The larvae are mainly found in very acid water, with pHs down to 3.5 (Schleuter, 1985; Moller Pillot, 2003). However, they are not absent from willow marshes and other marshy pools where the pH can be hardly below 7. Fittkau (1962:285) mentions reports by Goetghebuer in which the larvae were found in water at pHs ranging from 6.5 to 9. Leuven et al. (1987) found the species only at pHs lower than 5.5.

TROPHIC CONDITIONS AND SAPROBITY

The Dutch sites are often rich in nitrogen, even though this is probably no precondition for the larvae. Schleuter (1985) reports very different nitrogen contents at the sites where she found this species. Kreuzer (1940) found the larvae mainly in dystrophic water. Because the food probably consists of decomposing organic material, the characteristic environment is probably poor in most nutrients, but not oligosaprobic. The oxygen content at some sites described by Schleuter (1985: 25, 91) was very low during the day in summer, and so at night it may have dropped to nearly zero. It is not clear whether the larvae already had emerged at the time Schleuter made the measurements.

Telopelopia fascigera   (Verneaux, 1970)

SYSTEMATICS AND IDENTIFICATION

Telopelopia belongs to the *Thienemannimyia* group and is related to *Conchapelopia* (Murray and Fittkau, 1989). The larvae are easily distinguished from the other members of the group by the large basal tooth of the mandible (Fittkau and Roback, 1983).

DISTRIBUTION IN EUROPE

Very little is known about the distribution of *T. fascigera*. Caspers (1991) does not mention the species from the Rhine. However, Klink found the exuviae in the river Waal, a lower course of the Rhine in the Netherlands (Klink and Moller Pillot, 1996). It is probably a rare species of European rivers.

Thienemannimyia   Fittkau, 1957

SYSTEMATICS AND RELATIONSHIP

Thienemannimyia is closely related to *Arctopelopia, Conchapelopia* and *Rheopelopia* (Fittkau and Roback, 1983). In our key we use for these genera together the name *Conchapelopia* aggregate, although they do not form a monophyletic unit (Saether, 1977: 49). Many problems still have to be solved within the genus *Thienemannimyia* and some species names are still in doubt. At least a partial revision of the genus is needed (Spies and Saether, 2004).

EUROPEAN SPECIES

In Europe the genus includes at least 11 species, the adults of which cannot always be distinguished. At least *T. carnea* (Fabricius) and *T. pseudocarnea* Murray occur in the Netherlands. Old records of *Tanypus carneus* in the Netherlands refer to *Zavrelimyia nubila* (see De Meijere, 1939 and Fittkau, 1962: 300,301). The same is true for *Ablabesmyia carnea* in Goetghebuer (1911).

IDENTIFICATION

The adult male of *T. pseudocarnea* cannot be identified using most keys because the species has been described only in 1976 (Murray, 1976). Pupae and exuviae can be identified using Langton (1991). The larvae of most species have as yet not, or insufficiently, been described.

DISTRIBUTION IN EUROPE AND THE NETHERLANDS

The genus *Thienemannimyia* is widely distributed in Europe, but scarce in flat regions. For example, Serra-Tosio and Laville (1991) mention no species from north-western France. In the Netherlands exuviae of *T. carnea* and *T. pseudocarnea* have been found in the stretch of the Meuse that forms the border between Belgium and the Netherlands (Klink and Moller Pillot, 1996). Klink found a pupa of *T. pseudocarnea* in this stretch of the Meuse at Grevenbicht. A prepupa, probably belonging to a third species, has been found in the river Vecht just inside Germany before the border with the Netherlands. (Mulder, unpublished).

FEEDING AND BEHAVIOUR

The larvae creep around freely, preferably on a firm substrate. In the guts of 215 third and fourth instar larvae of *T. festiva*, Tokeshi (1991) found mainly remains of Chironomidae and to lesser extent Harpacticidae, Oligochaeta and other animals. The proportions of prey animals appeared to depend mainly on their density in the surrounding environment rather than the active preference of the predator. Algae and detritus formed a negligible part of the diet (at least in summer). Kawecka and Kownacki (1974) found fragments of Chironomidae and other animals in the guts of *T. geijskesi*.

MICROHABITAT

Pinder (1980) sampled larvae of *Thienemannimyia* in an English chalk stream exclusively on gravel. Pinder et al. (1987) found the larvae on the gravel as well as on water plants (*Ranunculus calcareus*), but not in soft sediment samples. Tokeshi (1991) sampled larvae of *T. festiva* on stones in a lake in Northern Ireland.

WATER TYPE

Current

The larvae of most of the species live in fairly fast-flowing streams. *T. pseudocarnea* and *T. carnea* may be expected more than other species to be present in lowland streams. The latter species inhabits mainly the salmonid region (Lehmann, 1971; Orendt, 2002), but occurs in mountainous regions especially downstreams (Laville and Vinçon, 1991; Michiels, 2004). *T. northumbrica* has been found in stagnant water bodies in hilly country (Fittkau, 1962: 190; confer Serra-Tosio and Laville, 1991: 44). *T. festiva* has been found also in lakes (Tokeshi, 1991).

Trissopelopia longimanus   (Staeger, 1839)

Pentaneura hieroglyphica Zavrel, 1936: 319

NOMENCLATURE AND RELATIONSHIPS

In the original publication by Staeger (1839) the epithet *longimanus* may be regarded as either a noun (manus = hand) or an adjective. According to the ICZN Code (1999: article 31.2.2) in such cases the species name is to be treated as a noun. *Trissopelopia* is related to *Pentaneura*, *Paramerina* and *Zavrelimyia* and resembles these genera especially as a pupa (Fittkau and Murray, 1986).

DISTRIBUTION IN EUROPE AND THE NETHERLANDS

T. longimanus is widely distributed in Europe but has not yet been recorded in all regions (Pankratova, 1977; Fittkau and Reiss, 1978). The related *T. flavida* is probably absent from the Netherlands and adjacent lowlands. In the Netherlands *T. longimanus* has been found only on the Veluwe and in central Limburg (Moller Pillot and Buskens, 1990; Nijboer and Verdonschot, unpublished data).

LIFE CYCLE

In the river Fulda and its tributaries exuviae and adults occur from May until November (Fittkau, 1962: 364). Lindegaard et al. (1975) found pupae or adults in May and August. They call the species bivoltine with a rapid development in summer months. The data of Hildrew et al. (1985) from an English stream correspond with this. In the Netherlands pupae have been sampled in May and June and fourth instar larvae from January until August. Third instar larvae are also present until April.

FEEDING

According to investigations by Hildrew et al. (1985) the second and third instar larvae feed mainly upon detritus, while older larvae are mainly carnivorous, at least in summer, consuming among others Chironomidae and juvenile Plecoptera. In winter the larvae appeared to be forced to change over to detritus, presumably involving a great reduction in food quality.

MICROHABITAT

In the Fulda region the larvae live in spring brooks with moderately flooded gravelly-sandy, but nevertheless silt-containing bottoms and in moss vegetation (Fittkau, 1962). In an English brook the larvae lived mainly between dead leaves and only in small numbers on a bottom of stones without leaves (Hildrew et al., 1985). Lindegaard et al. (1975) found the larvae rather numerous in the moss carpet of the Ravnkilde spring in Denmark. The larvae lived in the lowermost layer of the carpet, the detritus zone, composed of dead moss stems enclosed by deposits of rather coarse detrital particles.

DENSITIES

In an English brook the maximum numbers were probably approximately 200 larvae/m² during months with many second and third instar larvae (June and October) (Hildrew et al., 1985). In the moss carpet of the Ravnkilde spring the density varied from 92 to 1150 larvae/m² (Lindegaard et al., 1975). The Dutch records concern only few specimens at every site.

WATER TYPE

T. longimanus can be found in springs, mountain brooks and hygropetric environments (Zavrel, 1936; Fittkau, 1962; Lehmann, 1971; Lindegaard, 1995). In Scandinavia the larvae also live in lakes (Brundin, 1949; Raddum and Saether, 1981). In the Netherlands the species is not known from South Limburg, but it is found in the Veluwe region and central Limburg in rather fast flowing artificial groundwater creeks and brooks between 1 and 4 m wide, and their springs.

pH

Raddum and Saether (1981) found the larvae in two Norwegian lakes with pH 4.82 and 6.25. Lindegaard et al. (1975) caught *T. longimanus* in great numbers in the Ravnkilde spring (Denmark) at a pH of approx. 8. In the Netherlands the larvae were not found in the calcareous springs in the marl region (*Mergelland*) in southern part of the Dutch province of Limburg, but only in the less calcareous groundwater springs and creeks in central Limburg and the Veluwe region.

TROPHIC CONDITIONS

The springs and spring brooks in Denmark and the Netherlands are fed by groundwater. Lindegaard et al. (1975) mention a phosphate content of approx. 0.08 mg P/l and 3.01 to 6.45 mg NO_3-N/l. Moller Pillot and Krebs (1981) and Klink (1982a) attribute *T. longimanus* to a community of species living in poor groundwater. The high nitrate content found in the Danish spring shows that in overshadowed springs the presence of nutrients in itself does not have much influence on the system.

Xenopelopia   Fittkau, 1962

SYSTEMATICS AND RELATIONSHIP

Xenopelopia belongs to the tribe Pentaneurini within the subfamily Tanypodinae (Fittkau, 1962; Saether, 1977). Within this tribe the genus is most related to *Monopelopia* (Murray and Fittkau, 1989). The larva is very similar to the larva of *Telmatopelopia*.

EUROPEAN SPECIES

Two species of the genus occur in Europe: *X. falcigera* (Kieffer), 1911 and *X. nigricans* Fittkau, 1962. Because *X. nigricans* was not identified as a separate species before 1962, in the older literature the whole genus is named *Pelopia* (or *Ablabesmyia*) *falcigera*. Both species occur in the Netherlands.

IDENTIFICATION OF ADULTS AND PUPAE

The adult males can be identified using Fittkau (1962) and Pinder (1978). The colour is not a reliable characteristic, but according to Dettinger-Klemm (2003) the adults of the first spring generation are much darker than those of subsequent generations. The characteristics for distinguishing pupae and exuviae given by Pankratova (1977) and Langton (1991) are not reliable. In rearing *X. nigricans* we found even within one population a variation of the atrium from strongly sinuous to nearly straight. For this reason the numbers of sites where Nijboer and Verdonschot (2001) reported finding one or both species is doubtful. Dettinger-Klemm (2003: 293) came to the same conclusion and proposed studying other characteristics, such as the shape of the thorax comb.

DISTRIBUTION IN EUROPE AND THE NETHERLANDS

Xenopelopia occurs throughout nearly the whole of Europe, with the exception of the high Arctic and possibly some parts of the Mediterranean region (Fittkau and Reiss, 1978). In 1978 the distribution of the separate species was still underinvestigated. In the Netherlands the genus can be found in all regions (Moller Pillot and Buskens, 1990). Nijboer and Verdonschot (2001) call *Xenopelopia* very common in the light of records from all drainage areas. Based on 13 identifications of adult males, we found *X. falcigera* 3 times and *X. nigricans* 10 times. *X. falcigera* has been found in the Pleistocene areas and in the dunes. Near Bonn (Germany) Schleuter (1985) found *X. falcigera* to be much more common.

LIFE CYCLE

Pupae and adults occur from the end of March until October (in southern England from early March: Mundie, 1957). There are at least two generations a year; Dutch unpublished data and some data in the literature indicate the possibility of a third generation (Learner and Potter, 1974; Dettinger-Klemm and Bohle, 1996). Brundin (1949) found adults in Swedish lakes mainly from the end of April to early May. In autumn mainly juvenile larvae are found, and in winter third and fourth instar larvae are common. The scarcity of pupae in early autumn indicates a diapause, but this has not been proved. Further development of larvae may be blocked as early as August.

FEEDING AND BEHAVIOUR

Compared with most other Tanypodinae, the larvae of *Xenopelopia* are rather small: fourth instar larvae are up to 8 mm long. They creep around actively, moving their head in all directions. When threatened they move jerkily backwards. They seem to be uninterested in animal prey. From our own observations it appears that, very selectively, fine particles are often consumed (detritus rich in bacteria?). The guts have been found to contain mainly very fine detritus, sometimes also (mainly unicellular) algae and incidentally remains of (small?) animals. The gut is never totally filled, as in *Chironomus*. Most probably the basal food consists of detritus, bacteria, Protozoa and other periphyton (see the paragraph on feeding in the general introduction of the Tanypodinae). Specific literature on feeding and food of *Xenopelopia* is unknown to us.

MICROHABITAT

The larvae are nearly always found on a firm substrate and in places where they are protected from strong currents. They creep around, often in large numbers, between dead leaves on the bottom of pools and marshes. In ditches they are often most numerous near the banks or in dense vegetation. Although they seem to dislike swimming, and are probably very poor swimmers, they also live on the leaves of robust plants near the water surface, even on *Nymphaea* and *Nuphar* (water lilies) and sometimes also on *Lemna* (duckweed). In such cases densities are low. Reared larvae are observed on the bottom, on dead leaves and on plants and even at the water surface. The older larvae probably live nearer to the bottom than young larvae. At dusk the larvae are often caught in drift samples, possibly indicating that they move into less sheltered places to seek a favourable microenvironment.

DENSITIES

There are few quantitative data on density available and they are usually of low population densities. In early April 149 adults of *X. falcigera* emerged from 0.25 m² of a pool at the edge of a German forest and 96 emerged in June (Schleuter, 1985:127). In the Netherlands densities of more than 50 larvae/m² are found in ditches, pools, alder carr and a cut reed site in the littoral zone of a pond. Densities are always lower in larger or flowing water bodies.

WATER TYPE

Current

As a rule *Xenopelopia* is not recorded in fast-running brooks and streams, not even – after extensive investigations as by Lehmann (1971) in the Fulda and by Braukmann (1984) – in a great number of German streams. In the Netherlands many records of *Xenopelopia* are known from small and larger lowland brooks, dispersed over the whole country. However these always refer to very small numbers, except where the current was very low. The Roodloop at Hilvarenbeek was only temporarily colonised from stagnant ditches in the environment (Moller Pillot, 2003), and this may be the case in the majority of the lowland brooks. Verdonschot et al. (1992) mention the presence of the genus only in temporary upper courses.

Dimensions

According to the literature the larvae are common and sometimes numerous in small pools (Kreuzer, 1940; Schleuter, 1985) and more scarce in the littoral zone of lakes (Brundin, 1949; Fittkau, 1962). The vast majority of the Dutch sites are in small water bodies: pools, marshes, ditches and sometimes smaller upper courses of lowland brooks. In addition many recorded sites are in wider ditches and brooks and medium-sized or even large lakes. In larger or deeper water bodies the larvae seem to be confined to the littoral vegetation and numbers are usually low.

Permanence

Dettinger-Klemm and Bohle (1996) stated that some *Xenopelopia* adults emerged from a pool that consisted at that moment of just a moist bottom. Dutch investigations indicate that the larvae cannot survive when the soil has dried out completely. Kreuzer (1940) supposed that the eggs can survive in dry conditions. However, there is no indication of a true diapause, or even dormancy, in the egg stage. As far as can be ascertained, recorded sites in late summer have always been water bodies that have not recently completely dried out. Once this occurs, recolonisation seems to be necessary. *Xenopelopia* is certainly not characteristic of temporary water, although the larvae occur here frequently. Most probably the larvae exploit here the advantage of diminished competition and predation. The species can recolonise shortly after a period of drought because the flying period is long and ovipositing females probably select small water bodies (Dettinger-Klemm, 2003: 214).

TEMPERATURE

Fittkau (1962) supposes that the larvae of *X. nigricans* are thermophilous because they live in shallow stagnant water and do not occur in high northern parts of Europe. The fact that the adults can emerge as early as the end of March and in April, even in ditches with cold seepage water, seems to contradict this. However, it is possible that the eggs are deposited only in water at a relatively high temperature.

SOIL

Xenopelopia can be found in water bodies on every soil type. The Dutch data, gathered from the whole country, indicate a more frequent occurrence in sandy and peaty regions than in clay regions. In the province of Zeeland Tramper (1979) found no larvae in cattle pools, not even when these contained fresh water. Their scarce occurrence on clay was significant in water bodies in North Holland (Steenbergen, 1993). Rietveld et al. (1985) stated a preference for clay ditches in the province of Utrecht, but this may have been an accidental correlation with other factors. The scarce occurrence in Southern Limburg can be connected with the soil type or with scarcity of suitable habitats.

pH

Xenopelopia have rarely been found in very acid water. The larvae hardly occur in moorland pools, ditches and upper courses with a pH lower than 5.0 (Buskens, 1983; Verstegen, 1985; Leuven e.a., 1987; Duursema, 1996; Moller Pillot, 2003). However, there are diverse (often unpublished) records at lower pHs down to 3.8. Schleuter (185: 25/91) mentions the fairly numerous occurrence of *X. falcigera* and *X. nigricans* in a pool at a pH of 4.3. In such cases, the pH of the soil is most probably higher than in the water layer.

TROPHIC CONDITIONS AND SAPROBITY

Brundin (1949) found *Xenopelopia* only in more or less eutrophic lakes. According to Verdonschot et al. (1992) they occur most frequently in mesotrophic water bodies. In moorland pools and peat-bog lakes the larvae are found mainly after strong eutrophication, but this can be related to the pH (see Werkgroep Hydrobiologie MEC, 1993; Duursema, 1996). An analysis of 255 sites in Northern Holland (Steenbergen, 1993) indicates that the larvae occurred mainly in water with a relatively low phosphate content (up to 0.5 mg/l) and a low Kjeldahl nitrogen content (less than 2 mg N/l). They were also more common at lower oxygen content, and there was no correlation with ammonium content. In our own investigations in peaty ditches at Bergambacht, larvae were found at sites where hardly any oxygen was present at night and the ammonium nitrogen content could amount to more than 10 mg/l. For an inhabitant of pools with many decaying leaves one would expect the susceptibility for oxygen deficit and ammonia to be low. All the Dutch data indicate that the larvae are scarce in water affected by anthropogenic pollution. Their tolerance to low oxygen content can partly be attributed to the fact that the larvae are not true bottom inhabitants, but creep freely around on dead leaves and living water plants.

Both species of the genus may exhibit differences in their ecology. The water at all nine sites where *X. nigricans* has been found in the Netherlands is more or less eutrophic. *X. falcigera* was caught in a dune pool, a pool in a hayfield poor in nutrients and on a flat roof. However, both species can live together (Dettinger-Klemm and Bohle, 1996; Dettinger-Klemm, 2003).

SALINITY

In the Netherlands *Xenopelopia* is a pronounced inhabitant of fresh water. Krebs (1981, 1984, 1990) found never *X. nigricans* at chloride contents above 500 mg/l. Steenbergen (1993) mentions also some records at a chloride content above 1000 mg/l.

DISPERSAL

The frequent presence in temporary pools indicates much active or passive movement through the air. Dettinger-Klemm (2003) has reported frequent new colonisation by both species and counts them among the colonisers. It is unknown whether large distances are often covered. The larvae are often transported by drift, at least when they are common in pools or ditches in contact with the stream (Steinhart, 1999; Moller Pillot, 2003; cf. Schnabel, 1999).

Zavrelimyia Fittkau, 1962

SYSTEMATICS AND RELATIONSHIP

Within the Pentaneurini the genus *Zavrelimyia* is most closely related to *Paramerina* (Fittkau, 1962; Fittkau and Roback, 1983). The larva of *Schineriella*, which at first glance resembles *Zavrelimyia*, is not closely related (Murray and Fittkau, 1989).

EUROPEAN SPECIES

Five species of the genus have been reported in the Netherlands and adjacent lowlands (Fittkau, 1962; Ashe and Cranston, 1990; Langton, 1991; Lindegaard, 1995):

Zavrelimyia barbatipes (Kieffer, 1911)
Zavrelimyia hirtimana (Kieffer, 1918)
Zavrelimyia melanura (Meigen, 1804)
Zavrelimyia nubila (Meigen, 1830)
Zavrelimyia signatipennis (Kieffer, 1924).

At least three species occur in the Netherlands: *Z. barbatipes, Z. melanura* and *Z. nubila*. Only *Z. nubila* is a common species here and so the ecology of this species will be treated more extensively. The ecology of the other species seem to differ essentially from that of *Z. nubile* but some aspects are more characteristic for the genus as a whole and are treated under the genus.

IDENTIFICATION OF ADULTS

Identifying adults of this genus is very difficult and many mistakes have been made. For extensive descriptions and figures of all European species see Fittkau (1962). Identification of pupae and exuviae (using Langton, 1991) is more reliable.

MICROHABITAT

Tokeshi (1993) sometimes found *Zavrelimyia* larvae in great numbers on *Myriophyllum spicatum* in a lowland stream in England. Moller Pillot (2003) has also reported that the larvae of *Z. nubila* often remain on plants near the water surface. Moller Pillot and Buskens (1990: 10) therefore incorrectly state that this happens rarely. Elsewhere, however, there have been frequent reports of larvae staying on the bottom. Reared older larvae of *Z. nubila* usually creep on the bottom. According to Verdonschot and Schot (1987) larvae in helocrene springs usually live within the sediment. In forest brooks without vegetation Tolkamp (1980) rarely found the larvae on open sandy soil, but mainly where coarse detritus, such as dead leaves, were present. These were probably mainly *Z. barbatipes* larvae. Hildrew et al. (1985) also found this species mainly among dead leaves and almost never on stones. Schleuter (1985) observed in wheel ruts that *Z. barbatipes* (? see under) larvae could be numerous at sites with or without dense vegetation, dead leaves or both. It may be assumed that the larvae can more or less adapt themselves to the situation (however, see also under *Z. nubila*: Dispersal).

FEEDING AND BEHAVIOUR

The larvae creep freely around on the substrate and do not swim unless necessary. The fact that they are often found on fairly delicate or finely divided leaves indicates that they seek prey of small dimensions, as also Tokeshi (1993: 469) supposed. Hildrew et al. (1985) stated that larger larvae take larger prey. Second instar larvae of *Z. barbatipes* ate only detritus; in the diet of third and fourth instar larvae Chironomidae (mainly *Heterotrissocladius* and *Micropsectra*) and juvenile Plecoptera played a dominant role.

Reared *Z. nubila* larvae are easily frightened and catch mainly less mobile prey, but there seems to be no preference between Chironomidae, Oligochaeta and Cladocera (own observations).

pH

Orendt (1999) found larvae of the genus *Zavrelimyia* (mainly *Z. barbatipes?*) in German streams exclusively at a pH between 5.5 and 6.5. However, Braukmann (1984) found this species in carbonate rich streams as well. In the Netherlands *Zavrelimyia* larvae have rarely been caught at a pH lower than 5.0 (Moller Pillot and Buskens, 1990). Duursema (unpublished) found one larva in a moorland pool in Drenthe (pH approx. 4). Schleuter (1985) and Moller Pillot (2003) mention many records at pHs between 4.5 and 6 in wheel tracks in the upper course of a brook, whereas the majority of Dutch records refer to water bodies with a pH of 6 and more. Steenbergen, 1993 states that at pHs higher than 7.5 the genus is scarce in North Holland; this most probably refers only to *Z. nubila*. Larvae of *Z. nubila* and *Z. melanura* are sampled regularly in springs in the carbonate-rich hilly country of South Limburg, where the pH is expected to be more than 7.5.

Zavrelimyia barbatipes (Kieffer, 1911)

DISTRIBUTION IN EUROPE AND THE NETHERLANDS

Z. barbatipes has been found across almost the whole of Europe, but seems to be absent from Scandinavia (Fittkau and Reiss, 1978). In the Netherlands pupae and exuviae have been sampled in the sandy eastern and southern parts of the country.

LIFE CYCLE

In a southern English stream Hildrew et al. (1985) found mainly fourth instar larvae in June, mainly third instar larvae in July and third and fourth instar larvae in August. Their data indicate three generations a year. Lindegaard-Petersen (1972) found pupae in May and September and an adult female in July. In the Netherlands pupae and prepupae are caught mainly in April and May and in July, and (in northern Belgium) at the end of September. It is not clear whether the scarcity of records in summer and late summer is caused by a drop in the current that often occurs in small lowland brooks (compare however *Z. nubila*).

DENSITIES

Both in the Netherlands and in other countries *Z. barbatipes* seems to occur usually in low densities (Lehmann, 1971; Braukmann, 1984; compare Tokeshi, 1993). Lindegaard and Mortensen (1988) estimate 28 larvae/m² in a Danish brook. In a brook in southern England, however, the numbers most probably rose to 400–500/m², whereas the numbers in winter were not higher than 50/m² (Hildrew et al., 1985).

Themes not treated for this species can be found under the genus (above) or under Z. nubila (below).

WATER TYPE
Current

More than probably any other species within the genus, *Z. barbatipes* is an inhabitant of running water and hardly ever springs. In France the species is only known from the eastern and southern parts of the country and not from the flatter regions (Serra-Tosio and Laville, 1991). Lehmann (1971) found *Z. barbatipes* only in springs and mountain brooks and calls the species cold stenothermous. However, Braukmann (1984) found the species in lowland brooks. Hildrew et al. (1985) found *Z. barbatipes* in an English stream mostly where the current was very slow (less than 5 cm/s), at somewhat deeper places where dead leaves remained. In the Netherlands the species is confined to lowland brooks. Interestingly, Schleuter (1985) found *Z. barbatipes* fairly commonly in wheel tracks in a woodland region at a height of 170 m near Bonn in Germany. These mini-water-bodies had no connection with ground water, and *Z. nubila*, a characteristic inhabitant of small water bodies, was completely absent. These results may possibly be inaccurate because the adults of the different species are extremely hard to distinguish (Fittkau, 1962: 305). However, Dr. F. Reiss had checked the identification.

Dimensions
Z. barbatipes seems to be absent from large rivers.

Zavrelimyia hirtimana (Kieffer, 1918)

DISTRIBUTION IN EUROPE
Z. hirtimana has been found throughout nearly the whole of Central and Northern Europe, including Belgium, Germany, France and England (Fittkau, 1962; Samietz, 1999; Serra-Tosio and Laville (1991). The species may also be present in the Netherlands, but there are no reported data.

LIFE CYCLE
In an English chalk stream the adults emerged from February until November (Drake, 1982).

WATER TYPE
The literature on the ecology of *Z. hirtimana* exhibits remarkable differences. The type material was reared from a temporary woodland pool in the Czech Republic (Fittkau, 1962). Langton (1991) supposes that the species is a characteristic inhabitant of (nearly) stagnant pools in brooks. Drake (1982) caught fairly large numbers of emergent males from an English brook. According to Lindegaard (1995) in Germany the species is mainly known from limnocrenes but Orendt (1993) found the larvae only in a mesotrophic lake in Bavaria. *Z. hirtimana* is similar in its ecology to *Z. nubila*.

Zavrelimyia melanura (Meigen, 1804)

IDENTIFICATION
Z. melanura cannot be identified as a larva and is very difficult to identify as an adult (see Fittkau, 1962). Best of all the pupa can be identified (Langton, 1991). The larva has been described by Laville (1971).

DISTRIBUTION IN EUROPE AND THE NETHERLANDS
Z. melanura has been recorded across nearly the whole of Western and Central Europe (Fittkau and Reiss, 1978). In the Netherlands the species has been found only in Southern Limburg, in the Veluwe region (Higler et al., 1982) and near Winterswijk in the east. The record for the Maarsseveen lake by Kouwets and Davids (1984) is probably a mistake.

LIFE CYCLE
According to Lehmann (1971) in the Fulda *Z. melanura* probably has two generations, emerging in April/May and August/September. In the Netherlands pupae have been found in June, August and the end of September. The occurrence of three generations cannot be excluded (compare *Z. nubila*).

WATER TYPE

Z. *melanura* lives exclusively in springs, mountain brooks and alpine and northern lakes (Brundin, 1949 [as *Ablabesmyia nigropunctata*]; Fittkau, 1962; Lehmann, 1971; Braukmann, 1984; Langton, 1991). For further literature, see Lindegaard (1995). In the Netherlands the species has been found in spring brooks and the upper catchments of lowland brooks.

Zavrelimyia nubila   (Meigen, 1830)

IDENTIFICATION

Moller Pillot and Buskens (1990, p.21) erroneously have *nubila agg.* The pupae of Z. *nubila* and Z. *hirtimana* cannot be confused (see Langton, 1991, pp.48–50).

DISTRIBUTION IN EUROPE AND THE NETHERLANDS

Z. *nubila* has been recorded only in Western and Central Europe and in Russia (Fittkau and Reiss, 1978; Ashe and Cranston, 1990). In France Serra-Tosio and Laville (1991) know only sites in the south-eastern part of the country. In the Netherlands the species has been found nearly everywhere in the Pleistocene areas and in South Limburg. The records of *Zavrelimyia* larvae in the dunes and the peat marsh area of North Holland (Steenbergen, 1993; Moller Pillot and Buskens, 1990) most probably refer to this species, although in some cases the larvae or pupae can be confound with those of *Schineriella*. There are no reported catches from the Wadden Islands and the Holocene clay regions.

LIFE CYCLE

Some data in the literature indicate two generations a year (Tokeshi, 1993; Dettinger-Klemm, 2003: 19/20, 80), but three generations are also possible, as Orendt (1993) found most probably for Z. *hirtimana* in Bavaria. In the Netherlands we found strong indications for three generations a year, from the end of March to the end of September, at least where temperatures can be high. There are many more records of pupae in spring than in late summer and the absence of pupae in autumn and winter indicates an autumnal diapause. From the end of October until end of March larvae in third and fourth instar are found together.

FEEDING

See under the genus.

MICROHABITAT

See under the genus.

DENSITIES

Densities of larvae are usually very low and stream samples often only contain just one or more larvae. In pools, however, densities of more than 100 larvae/m^2 can often be found. An extremely high density was reported on 15 June 1992 in the Roodloop at Hilvarenbeek, at that time only one metre wide and with a closed vegetation of mainly grasses. A sample of 0.15 m^2 contained 92 larvae and 12 pupae, also nearly 700 specimens/m^2 (see Moller Pillot, 2003).

WATER TYPE

Current and dimensions

Fittkau (1962) calls the species a typical inhabitant of different types of pools. In the Netherlands Z. *nubila* lives in small water bodies of very different types: ditches, springs, marshes with seepage water, lowland brooks and moorland pools. There are even records from flat roofs and from a concrete tank in a garden. Buskens (unpublished) found the species only once in a larger sand pit.

Permanence
The species is rather common in temporary water, but cannot survive complete desiccation of both water and soil.

TEMPERATURE

In contrast to most other species of the genus, *Z. nubila* seems to be thermophilous (Fittkau, 1962), although its presence in springs and seepage water indicates an independence of water temperature.

SOIL

In the Netherlands there are no reported sites on clay, but Fittkau (1962) mentions the species as being present in a clay pit in Germany.

pH

See under the genus.

TROPHIC CONDITIONS AND SAPROBITY

The presence of *Z. nubila* seems to be independent of the trophic or saprobic situation. Low oxygen contents present no problems. High densities can be favoured by the presence of much animal food, usually owing to the presence of decomposing material.

SALINITY

Fittkau (1962) mentions the presence of *Z. nubila* in brackish water. In the Netherlands, however, no records are known from water with more than 300 mg Cl/l.

DISPERSAL

Because *Z. nubila* is often present in temporary water without surviving after complete desiccation (of water and bottom sediment), it is clear that these places are easily colonised. Dettinger-Klemm (2003: 74, 214) also counts the species among the 'colonisers'. It is unknown whether *Z. nubila* has developed qualities specially useful for this purpose, such as quickened development of larvae or higher flying by females for dispersal over greater distances. Moller Pillot (2003) found the larvae in the Roodloop fairly often in drift samples, but only in winter. This species of stagnant water is possibly more than related species sensible to downstream transport, when the larvae are creeping on plants near the water surface. Downstream transport from more stagnant water may also be the reason for the great numbers of *Zavrelimyia* found by Tokeshi (1993) in spring in the river Tud.

Zavrelimyia signatipennis   (Kieffer, 1924)

DISTRIBUTION IN EUROPE

Z. signatipennis has been found in the Alps and northern Germany (Fittkau, 1962). It is probably not present in the Netherlands.

WATER TYPE

The larva is a characteristic inhabitant of springs, and in central Germany also in mountain streams (Fittkau, 1962; Lehmann, 1971; Lindegaard, 1995).

(B) TABLES OF BIOLOGICAL AND ECOLOGICAL PROPERTIES OF TANYPODINAE LARVAE
(in principle for the Dutch situation)

In tables 2 to 4 valence values are given as in Moog (1995). The sum of the values for one factor is always 10. Rare exceptions are not mentioned (for instance, larvae which cannot survive in such an environment). For all tables: blank = unknown; ? = not sure.

TABLE 1: General biology

gener	=	number of generations/yr
adult	=	flying period of adults (and egg deposition): months
developm	=	duration of larval development in months
		(if more generations: in summer)
diap	=	diapause: in summer or in winter
hibern	=	hibernation: larval instar
eggs	=	number of eggs in one egg mass
food	=	only the most important food for third and fourth instar is given. Younger larvae eat more detritus and other fine materials. AN = animals, AL = algae, FP = fine particulate organic material (in the case of Tanypodinae, probably selectively rich in bacteria, etc.).
habitat	=	preferred microhabitat of the larvae: bo = near or in the bottom; pl = on plants; surface = near the surface (usually on plants)

	gener	adult	developm	diap	hibern	eggs[2]	food	habitat
Ablabesmyia longistyla	3?	4–10	2?	winter	2/3		AN	bo, pl
Ablabesmyia monilis	3	4–10	2	winter	2/3	320	AN	bo, pl
Ablabesmyia phatta	2(3?)	4–10	2?	winter	2/3		AN	bo, pl
Anatopynia plumipes	1	2–4	12	winter	4	700	AN	bo
Apsectrotanypus trifasc.	2(3?)	4–10	3	winter	3/4	420	AN	bo
Arctopelopia barbitarsis	?				3/4?		AN?	bo?
Clinotanypus nervosus	1	6(-7)	12	winter	4(3)	370	AN	bo
Conchapelopia melanops	2	4–9	3		3/4	350	AN	bo, pl
Guttipelopia guttipennis		5–8					AN?	pl
Krenopelopia	2	5–8	3		3/4			bo
Labrundinia longipalpis						(127)	AN?	
Macropelopia adaucta	2(3?)	3–10	3		2–4		AN?	
Macropelopia nebulosa	2(3?)	3–10	3		2–4	320	AN?	bo
Macropelopia notata	2	4–10	3		2–4?		AN?	
Monopelopia tenuicalcar	2	5–9	3	winter	3(4)		FP?	surface
Natarsia	2?	5–8	3?		4(3?)		AN/AL	
Paramerina cingulata	3?	4–9	2–3		2–4			pl
Procladius (Holotanypus)	2	4–10[3]	3	winter		350	AN	bo
Procladius (Psilotanypus)	(1-)2	4–8	2?	winter		450	AN	bo
Psectrotanypus varius	3	3–10	2		3/4	320	AN	bo
Rheopelopia							AN?	
Tanypus kraatzi	2	5–10	3	winter	2–3(-4)	600	AL	bo
Tanypus punctipennis	2	5–10	3	winter	2–4?	800	AL	bo
Tanypus vilipennis	2?		3?		2–4?	600	AL	bo
Telmatopelopia nemorum	1(2?)	4–7	12?	summer	3/4		FP	bo
Thienemannimyia							AN	bo, pl
Trissopelopia longimanus	2	5–10			3/4		AN	bo
Xenopelopia	2–3	4–10	2	winter?	3/4		FP	bo, pl
Zavrelimyia barbatipes	3	4–9	3	winter?	3/4	(245)	AN	bo, pl
Zavrelimyia melanura	2?	4–9	3	winter?	3/4	(245)	AN	bo, pl
Zavrelimyia nubila	3	4–9	2	winter?	3/4	(245)	AN	bo, pl

[2] Eggs: if bold: mean of two or more data. Nearly all data from Koreneva (1959) and Nolte (1993). According to Dettinger-Klemm, egg numbers are temperature dependent. Higher numbers can be expected at lower temperatures.
[3] In basins for drinking water as late as December.

TABLE 2: Saprobity and oxygen

For the definition of saprobity we follow Sladecek (1973: 28): the amount and intensity of decomposition of organic matter. Moog (1995) defines saprobic valence as the tolerance of organic substances, primarily determined by the availability of organic food and oxygen. The differences between our figures and those of Moog are mainly attributable to the fact that we separate the direct influence of organic material as far as possible from the influence of oxygen content. Moreover, Moog's figures concern Austrian flowing waters and our figures are for all types of stagnant and flowing waters. The Netherlands has many types of water bodies which are rare or absent in Austria. *Monopelopia*, *Xenopelopia* and *Zavrelimyia* are considered oligosaprobic by Moog, but in the Netherlands these larvae often live in very polluted stagnant water. For further information, see sections 2.6.4 and 2.6.6.

ol	= oligosaprobic
B	= beta-mesosaprobic
A	= alpha-mesosaprobic
p	= polysaprobic
stab	= stable oxygen regime: always above 50% saturation
unst	= unstable: minimum between 10 and 50% saturation
low	= sometimes (but not longer than a few hours) less than 5% saturation
rott	= rotting: in summer almost daily less than 5% saturation for hours

	saprobity							oxygen			
	ol	ol/B	B	B/A	A	A/p	p	stab	unst	low	rott
Ablabesmyia longistyla	1	2	2	2	2	1	0	4	4	2	0
Ablabesmyia monilis	1	2	2	2	2	1	0	4	4	2	0
Ablabesmyia phatta	2	2	2	2	1	1	0	5	4	1	0
Anatopynia plumipes	0	0	2	3	3	2	0	1	4	3	2
Apsectrotanypus trifasc.	0	1	3	3	2	1	0	5	5	0	0
Arctopelopia barbitarsis								4	5	1	0
Clinotanypus nervosus	0	1	4	3	2	0	0	3	5	2	0
Conchapelopia melanops	0	1	3	3	2	1	0	4	5	1	0
Guttipelopia guttipennis	1	3	3	2	1	0	0	4	4	2	0
Krenopelopia[4]	3	3	3	1	0	0	0	10	0	0	0
Labrundinia longipalpis	3	3	3	1	0	0	0	7	3	0	0
Macropelopia adaucta	1	2	3	2	1	1	0	3	5	2	0
Macropelopia nebulosa	0.5	1	2	3	2	1	0.5	3	5	2	0
Macropelopia notata								8	2	0	0
Monopelopia tenuicalcar	0	1	2	2	2	2	1	2	3	3	2
Natarsia	1	1	2	2	2	2	0	3	3	3	15
Paramerina cingulata	1	3	3	2	1	0	0	5	4	1	0
Procladius (Holotanypus)	1	1	2	2	2	1	1	2	5	2	1
Procladius (Psilotanypus)											
Psectrotanypus varius	0	0	1	2	2.5	2.5	2	1	3	4	2
Rheopelopia	0	1	2	3	3	1	0	5	5	0	0
Tanypus kraatzi	0	0	1	2	3	2	2	2	3	3	2
Tanypus punctipennis	0	0	1	2	3	2	2	2	3	3	2
Tanypus vilipennis	0	2	3	3	2	0	0	4	5	1	0
Telmatopelopia nemorum	0	1	2	2	2	2	1	2	3	3	2
Thienemannimyia											
Trissopelopia longimanus	3	4	3	0	0	0	0	8	2	0	0
Xenopelopia	0	1	2	2	2	2	1	1	3	4	2
Zavrelimyia barbatipes	0	1	4	4	1	0	0	5	5	0	0
Zavrelimyia melanura								10	0	0	0
Zavrelimyia nubila	0	1	2	2	3	2	0	3	3	3	1

TABLE 3: pH and salinity

The pH mentioned is the pH of the water column! (In the soil the pH is often more neutral.)
Salinity in the often silty soil can be much higher than in the water layer (see section 2.6.8).

	pH					salinity (g Cl/l)				
	< 4.5	5	6	7	>7.5	<0.3	0.3–1	1–3	3–10	>10
Ablabesmyia longistyla	2	2	2	2	2	9	1	0	0	0
Ablabesmyia monilis	2	2	2	2	2	9	1	0	0	0
Ablabesmyia phatta	2	2	2	2	2	9	1	0	0	0
Anatopynia plumipes	0	2	2	3	3	9	1	0	0	0
Apsectrotanypus trifasc.	0	0	2	4	4	10	0	0	0	0
Arctopelopia barbitarsis	0	0	3	4	3	10	0	0	0	0
Clinotanypus nervosus	0	0	2	4	4	8	2	0	0	0
Conchapelopia melanops	0	1	3	3	3	10	0	0	0	0
Guttipelopia guttipennis	2	2	2	2	2	10	0	0	0	0
Krenopelopia[4]	0	1	1	3	5	10	0	0	0	0
Labrundinia longipalpis	0	0	2	4	4	10	0	0	0	0
Macropelopia adaucta	3	3	3	1	0	10	0	0	0	0
Macropelopia nebulosa	1	2	3	3	1	9	1	0	0	0
Macropelopia notata[5]	1	1	1	3	4	10	0	0	0	0
Monopelopia tenuicalcar	2	2	2	2	2	9	1	0	0	0
Natarsia[6]	2	2	2	2	2	10	0	0	0	0
Paramerina cingulata	0	0	1	4	5	10	0	0	0	0
Procladius (Holotanypus)	1	2	2	2.5	2.5	4	3	2	1	0
Procladius (Psilotanypus)	0	1	3	3	3					
Psectrotanypus varius	1	1	2	3	3	6	3	1	0	0
Rheopelopia	0	0	0	5	5	10	0	0	0	0
Tanypus kraatzi	0	0	1	2	7	5	4	1	0	0
Tanypus punctipennis	0	0	0	2	8	3	3	3	1	0
Tanypus vilipennis	0	0	1	4	5	10	0	0	0	0
Telmatopelopia nemorum	3	4	2	1	0	10	0	0	0	0
Thienemannimyia						10	0	0	0	0
Trissopelopia longimanus	0	0	4	4	2[7]	10	0	0	0	0
Xenopelopia	0	1	3	4	2	8	2	0	0	0
Zavrelimyia barbatipes	0	2	3	4	1	10	0	0	0	0
Zavrelimyia melanura	0	1	3	3	3	10	0	0	0	0
Zavrelimyia nubila	0	2	3	3	2	9	1	0	0	0

[4] Krenopelopia often live above the water level, independent of water quality.
[5] *Macropelopia notata* is not found in the Netherlands in acid waters, but can live there according to Orendt (1999).
[6] Natarsia larvae can live around or above water level if oxygen levels are low (see Krenopelopia).
[7] *T. longimanus* has not been found in the Netherlands at high pH, but is able to endure it (see text).

TABLE 4: Current and permanence

(Species given a 0 for stronger current are usually carried along by currents).
A water body is called dry when also the bottom is dry.

	current (cm/s)					permanence (dry weeks/y.)				
	< 5	5–10	10–15	15–25	> 25	> 12	6–12	< 6	rarely	not
Ablabesmyia longistyla	6	2	1	1	0	0	1	1	2	6
Ablabesmyia monilis	6	2	1	1	0	0	1	1	2	6
Ablabesmyia phatta	9	1	0	0	0	0	1	1	3	5
Anatopynia plumipes	10	0	0	0	0	0	0	0	2	8
Apsectrotanypus trifasc.	0	1	2	3	4	0	0.5	1	2	6.5
Arctopelopia barbitarsis	0	3	4	3	0	0	0	0	0	10
Clinotanypus nervosus	6	3	1	0	0	0	0	0	1	9
Conchapelopia melanops	1	1	2	3	3	0	1	1	2	6
Guttipelopia guttipennis	10	0	0	0	0	0	1	1	2	6
Krenopelopia*	8	0.5	0.5	0.5	0.5	2	2	2	2	2
Labrundinia longipalpis	10	0	0	0	0	0	0	0	0	10
Macropelopia adaucta	3	3	3	1	0	1.5	1.5	2	3	2
Macropelopia nebulosa	1	2	3	3	1	0.5	0.5	1	4	4
Macropelopia notata**	8	2	0	0	0					
Monopelopia tenuicalcar	10	0	0	0	0	0	1	1	2	6
Natarsia	3*	3	3	1	0	1	1	1	3	4
Paramerina cingulata	7	2	1	0	0	0	0	0	2	8
Procladius (Holotanypus)	3	3	2	1	1	1	1	1	2	5
Procladius (Psilotanypus)	10	0	0	0	0	0	0	0	0	10
Psectrotanypus varius	4	3	2	1	0	1	1	1	4	3
Rheopelopia	0	1	3	3	3	0	0	0	0	10
Tanypus kraatzi	8	2	0	0	0	0	0.5	0.5	1	8
Tanypus punctipennis	9	1	0	0	0	0	0.5	0.5	1	8
Tanypus vilipennis	10	0	0	0	0	0	0	0	0	10
Telmatopelopia nemorum	6	3	1	0	0	2.5	2.5	2.5	1.5	1
Thienemannimyia	0	0	3	3	4	0	0	0	0	10
Trissopelopia longimanus										
Xenopelopia	8	2	0	0	0	1.5	1.5	1.5	3	2.5
Zavrelimyia barbatipes	0	0	3	3	4	0	1	1	2	6
Zavrelimyia melanura**	8	2	0	0	0					
Zavrelimyia nubila	7	2	1	0	0	2	2	2	2	2

* seepage water
** springs

REFERENCES

Aagaard, K., 1974. Morphological changes caused by nematode parasitism in Tanypodinae Diptera, Chironomidae). - Norsk ent. Tidsskr. 21: 11-14.

Aagaard, K. & B. Sivertsen, 1980. The benthos of Lake Huddingsvatn, Norway, after five years of mining activity. - In: Murray, D.A. (ed.): Chironomidae. Ecology, systematics, cytology and physiology. Oxford: 247-254.

Aleksevnina, M.S. & A.I. Bakanov, 1983. The space distribution of larvae and their migration. - In: N. Yu. Sokolova (ed.): Chironomus plumosus L. (Diptera, Chironomidae): 189-200. Moscow, Nauka. (In Russian)

Aleksevnina, M.S. & N. Yu. Sokolova, 1983. Multiplication, development and life cycle. - In: N. Yu. Sokolova (ed.): Chironomus plumosus L. (Diptera, Chironomidae): 156-188. Moscow, Nauka. (In Russian)

Armitage, P.D., 1968. Some notes on the food of the chironomid larvae of a shallow woodland lake in South Finland. - Ann. Zool. Fenn. 5: 6-13.

Armitage, P.D., P.S.Cranston & L.C.V. Pinder, 1995. The Chironomidae - Biology and ecology of non-biting midges. - Chapman & Hall (London).

Ashe, P., 1983. A catalogue of chironomid genera and subgenera of the worldincluding synonyms (Diptera: Chironomidae). - Ent. scand. Suppl. 20: 1-68.

Ashe, P. & P.S. Cranston, 1990. Family Chironomidae. - In: Soos, A. & L. Papp (eds.): Catalogue of Palaearctic Diptera 2. Akadémiai Kiadó, Budapest: 113-355.

Baker, A.S. & A.J. McLachlan, 1979. Food preferences of Tanypodinae larvae (Diptera: Chironomidae). - Hydrobiologia 62: 283-288.

Balushkina, E.V., 1987. Functional importance of the larvae of chironomids in continental water bodies. - Trudy zool. inst. Akad. Nauk SSSR 142: 1-179. (in Russian).

Baz', L.G., 1959. Biologiya i morfologiya predstavitelej roda Microtendipes, obitayushchikh v vodoprovodnom kanale Uchinskogo vodokhranilishche. - Trudy vses.gidrobiol. Obshch. 9: 74-84.

Beattie, D.M., 1978. Chironomid populations in the Tjeukemeer. - Thesis Leiden Univ. 150 pp.

Bell, H.L., 1970. Effects of pH on the life cycle of the midge Tanytarsus dissimilis. - Can. Ent. 102: 636-639.

Belyavskaya, L.I., 1956. The feeding of the larvae of Anatopynia varia F. - Trudy Saratov. otd. VNIORCh 4: 192-197. (in Russian).

Belyavskaya, L.I. & A.S. Konstantinov, 1956. The feeding of the larvae of Procladius choreus Meig. (Chironomidae, Diptera) and the loss they cause to the food base of fishes. - Vopr. Ikhtiol. 7: 193-203. (in Russian)

Biever, K.D., 1971. Effect of diet and competition in laboratory rearing of chironomid midges. - Ann. ent. Soc. Am. 64: 1166-1169.

Bijlmakers, L., 1983. De verspreiding en oecologie van chironomidelarven (Chironomidae: Diptera) in twee vennen in de omgeving van Oisterwijk (N.Br.). - Verslag K.U. Nijmegen. 118 pp. + bijl.

Bilyj, B., 1988. A taxonomic review of Guttipelopia (Diptera: Chironomidae). - Ent. scand. 19: 1-26.

Borutskij, E.V., 1963. Emergence of Chironomidae (Diptera) from continental water-bodies of different climatic belts as a factor of food supply for fishes. - Zool. Zh. 42: 233-247.

Branch, H.E., 1923. The life history of Chironomus cristatus Fabr. with descriptions of the species. - J. New York ent. Soc. 31: 15-30, Pl. I-V.

Braukmann, U., 1984. Biologischer Beitrag zu einer allgemeinen regionalen Bachtypologie. - Thesis Giessen Univ. 473 pp.

Brenner, R.J., M.J. Wargo, G.S. Stains & M.S. Mulla, 1984. The dispersal of Culicoides mohave (Diptera: Ceratopogonidae) in the desert of southern California. - Mosquito News 44: 343-350.

Brinkhof, M.W.G., 1995. Timing of reproduction. An experimental study in coots. - Thesis Groningen Univ. 160 pp.

Brock, T.C.M., 1985. Ecological studies on nymphaeid water plants. - Thesis Nijmegen Univ. 204 pp.

Brundin, L., 1949. Chironomiden und andere Bodentiere der südschwedischen Urgebirgsseen. - Rep. Inst. Freshw. Res. Drottningholm 30: 1-914.

Buskens, R., 1983. De makrofauna, in het bijzonder de chironomiden, en de vegetatie van een vijftigtal geëutrofieerde, zure of laag-alkaliene stilstaande wateren op de Nederlandse zandgronden. - Lab. Aquat. Oecol., Nijmegen, Report 159: 1-72 + app.

Buskens, R..F.M., 1987. The chironomid assemblages in shallow lentic waters differing in acidity, buffering capacity and trophic level in the Netherlands. - Ent. scand. Suppl. 29: 217-224.

Buskens, R.F.M. & G.A.M. Verwijmeren, 1989. The chironomid communities of deep sand pits in the Netherlands. - Acta Biol. Debr. Oecol. Hung. 3: 51-60.

Carpentier, C.J., A.M.J.P. Kuijpers, H.A.M. Ketelaars & F.E. Lambregts-van de Clundert, 1999. Veranderde macro-invertebratensamenstelling en -dichtheid als gevolg van de ontharding met calciumhydroxide in het spaarbekken Petrusplaat. - Water Storage Corporation Brabantse Biesbosch Ltd, Werkendam. 41 pp.

Casas, J.J. & A. Vilchez-Quero, 1989. A faunistic study of the lotic chironomids (Diptera) of the Sierra Nevada (S.E. of Spain): changes in the structure and composition of the populations between spring and summer. - Acta Biol. Debr. Oecol. Hung. 3: 83-94.

Caspers, N. 1980. Die Makrozoobenthos-Gesellschaften des Rheins bei Bonn. - Decheniana (Bonn) 133: 93-106.

Caspers, N., 1991. The actual biocoenotic zonation of the river Rhine exemplified by the chironomid midges (Insecta, Diptera). - Verh. Internat. Verein. Limnol. 24: 1829-1834.

Chamier, A.-C., 1987. Effect of pH on microbial degradation of leaf litter in seven streams of the English Lake District. - Oecologia (Berlin) 71: 491-500.

Chandler, P., 1998. Checklist of Insects of the British Isles (New series). Part 1: Diptera. - Handb. Ident. Br. Insects 12: I-XX, 1-234.

Cunningham-van Someren, G.R., 1975. A further note on swarming of male mosquitoes and other Nematocera in Kenya. - Entomologist's mon. Mag. 111: 147-160.

Danks, H.V., 1971. Life history and biology of Einfeldia synchrona (Diptera: Chironomidae). - Can. Ent. 103: 1597-1606.

Danks, H.V., 1971a. Overwintering of some north temperate and arctic Chironomidae. II. Chironomid biology. - Can. Ent. 103: 1875-1910.

Davies, B.R., 1976. The dispersal of Chironomidae larvae: a review. - J. ent. Soc. sth. Africa 39: 39-62.

Davies, L.J. & H.A.Hawkes, 1981. Some effects of organic pollution on the distribution and seasonal incidence of Chironomidae in riffles in the river Cole. - Freshw. Biol. 11: 549-559.

De Meijere, J.C.H., 1939. Naamlijst van Nederlandsche Diptera afgesloten 1 april 1939. - Tijdschr. Ent. 82: 137-174.

Delettre, Y., P. Tréhen & P. Grootaert, 1992. Space hetero-geneity, space use and short-range dispersal in Diptera: A case study. – Landsc. Ecol. 6: 175-181.

Delettre, Y.R., 1988. Chironomid wing length, dispersal ability and habitat predictability. – Holarctic Ecol. 11: 166-170.

Delettre, Y.R., 1989. Influence de la durée et de l'intensité de l'assèchement sur l'abondance et la phénologie des Chironomides (Diptera) d'une mare semi-permanente peu profonde. – Arch. Hydrobiol. 114: 383-399.

Dettinger-Klemm, P.-M. A., 2003. Chironomids (Diptera, Nematocera) of temporary pools – an ecological case-study. – Thesis Marburg Univ. 371 pp.

Dettinger-Klemm, P.-M. A. & H.W. Bohle, 1996. Über-lebungsstrategien und Faunistik von Chironomiden (Chironomidae, Diptera) temporärer Tümpel. – Limnologica 28: 403-421.

Downes, J.A., 1969. The swarming and mating flight of Diptera. – Ann. Rev. Ent. 14: 271-298.

Downes, J.A., 1974. The feeding habits of adult Chironomidae. – Ent. Tidskr. Suppl.95: 84-90.

Drake, C.M., 1982. Seasonal dynamics of Chironomidae (Diptera) on het Bulrush Schoenoplectus lacustris in a chalk stream. – Freshw. Biol. 12: 225-240.

Dusoge, K.,1980. The occurrence and role of the predatory larvae of Procladius Skuse (Chironomidae, Diptera) in the benthos of lake Sniardwy. – Ekol. Polska 28: 155-186.

Duursema, G., 1996. Vennen in Drenthe. Een onderzoek naar ecologie en natuur op basis van macrofauna. – Assen, Zuiveringsschap Drenthe. 140 pp.

Edwards, F.W., 1929. British non-biting midges (Diptera, Chironomidae). – Trans. ent. Soc. Lond. 77: 279- 430.

Egglishaw, H.J., 1968. The quantitative relationship between bottom fauna and plant detritus in strams of different calcium concentrations. – J. appl. Ecol. 5: 731-740.

Ferrarese, U., 1983. Chironomidi, 3 (Diptera: Chironomidae: Tanypodinae). – Guide per il riconosci-mento delle specie animali delle acque interne Italiane 26: 1-67. Verona.

Fittkau, E.J., 1962. Die Tanypodinae (Diptera Chironomidae). – Abh. Larvalsyst. Insekten 6: 1-453.

Fittkau, E.J. & D.A. Murray, 1986. The pupae of Tanypodinae (Diptera: Chironomidae) of the Holarctic region Keys and diagnoses. – Ent. scand. Suppl. 28: 31-113.

Fittkau, E.J. & F. Reiss, 1978. Chironomidae. – In: Illies, J. (ed.): Limnofauna europaea. 2. Aufl. Stuttgart: 404-440.

Fittkau, E.J. & S.S. Roback, 1983. The larvae of Tanypodinae (Diptera: Chironomidae) of the Holarctic region – Keys and diagnoses. – Ent. scand. Suppl. 19: 33-110.

Fritz, H.-G., 1981. Über die Mückenfauna eines tem-porären Stechmückenbrutgewässers des Naturschutzgebietes "Kühkopf-Knoblochsaue". – Hessische Faun. Briefe 1: 37-58.

Gibson, N.H.E., 1945. On the mating swarms of certain Chironomidae (Diptera). – Trans. R. ent. Soc. Lond. 95: 263-294.

Gijsen, M.E.A. & T.H.L. Claassen, 1978. Integraal structu-urplan Noorden des lands, landsdelig milieu-onder-zoek. Deelrapp. 2: Biologisch wateronderzoek: macro-fyten en macrofauna. – Rijksinst. Natuurbeheer, Leersum: 1-121 + bijl.

Glick, P.A., 1939. The distribution of insects, spiders and mites in the air. – Tech. Bull. U.S. Dept. Agric. 763. 151 pp.

Goddeeris, B., 1983. Het soortspecifieke patroon in de jaarcyclus van de Chironomidae (Diptera) in twee visvijvers te Mirwart (Ardennen). – Thesis Kath. Univ. Leuven. 177 pp. + bijl.

Goddeeris, B.R.,1987. The time factor in the niche space of Tanytarsus-species in two ponds of the Belgian Ardennes (Diptera Chironomidae). – Ent. Scand. Suppl. 29: 281-288.

Goddeeris, B.R.,1989. A methodology for the study of the life cycle of aquatic Chironomidae (Diptera). – Verh. Symposium Invertebr. België 1989: 379-385.

Goddeeris, B.R., 1990. Life cycle characteristics in Tanytarsus sylvaticus (van der Wulp, 1859) (Chironomidae, Diptera). – Ann. Limnol. 26: 51-64.

Goddeeris, B.R., A.C. Vermeulen, E. De Geest, H. Jacobs, B. Baert & F. Ollevier, 2001. Diapause induction in the third and fourth instar of Chironomus riparius (Diptera) from Belgian lowland brooks. – Arch. Hydrobiol. 150: 307-327.

Goetghebuer, M., 1911. Chironomides de Belgique. – Ann. Soc. ent. Belg. 15: 95-113.

Goetghebuer, M., 1923. Nouveaux matériaux pour l'étude de la faune des Chironomides de Belgique. – Ann. Biol. Lacustre 12: 103-120.

Goetghebuer, M., 1934. Catalogue des Chironomides de Belgique. – Ann. Soc. Ent. Belg. 74: 209-213.

Goetghebuer, M., 1936. Tendipedidae (Chironomidae). a) Subfamilie Pelopiinae (Tanypodinae). A. Die Imagines. – In: Lindner, E. (ed.): Die Fliegen der palaearktischen Region 13b: 1-50.

Gouin, F.J., 1959. Morphology of the larval head of some Chironomidae (Diptera, Nematocera). – Smithson. misc. Collns 137: 175-201.

Graaf, E. de, 1983. Verspreiding van chironomiden in relatie met miliefaktoren in de omgeving van Wageningen. – Ongepubl. rapport Natuurbeheer LH Wageningen 695: 85 + 126 pp.

Grodhaus, G., 1971. Sporadic parthenogenesis in three species of Chironomus (Diptera). – Can. Ent. 103: 338-340.

Grodhaus, G., 1980. Aestivating chironomid larvae asso-ciated with vernal pools. – In: Murray, D.A. (ed.): Chironomidae: ecology, systematics, cytology and physiology: 315-322. Oxford, Pergamon Press.

Hall, R.E., 1951. Comparative observations on the chi-ronomid fauna of a chalk stream and a system of acid streams. – J. Soc. Br. Ent. 3: 253-262.

Hamilton, A.L., 1965. An analysis of a freshwater benthic community with special reference to the Chironomidae. – Thesis dept. Zool. Univ. British Columbia. 93 + 216 pp.

Hammen, H. van der, 1992. De macrofauna van Noord-Holland. – Prov. Noord-Holland, Haarlem. 256 pp.

Hayashi, F.,1990. Factors affecting the body size at matu-ration of aquatic insects. – Japanese J. Limnol. 51: 199-215 (in Japanese).

Heinis, F. & W.R. Swain,1986. Impedance conversion as a method of research for assessing behavioral responses of aquatic invertebrates. – Hydrobiol. Bull. 19: 183-192.

Heinis, F., 1993. Oxygen as a factor controlling occurrence and distribution of chironomid larvae. – Thesis Amsterdam Univ. 155 pp.

Heinis, F., K.R. Timmermans & W.R. Swain, 1990. Short-term sublethal effects of cadmium on the filter feeding chironomid larva Glyptotendipes pallens (Meigen) Diptera. – Aq. Toxicol. 16: 73-86.

Heinis, F., W.J. van de Bund & C. Davids, 1989. Avoidance of low oxygen and food concentrations by the larvae of Tanytarsus species. – Acta Biol. Debr. Oecol. Hung. 3: 141-145.

Hermans, T. & M. Kahmann, 1985. Hydrobiologische inventarisatie van de Eempolders en het noordelijk gedeelte van de Utrechtse heuvelrug. – Prov. Waterstaat Utrecht, afd. ecologie. Rapport 54. 59 pp. + bijl.

Higler, L.W.G., 1977. Macrofauna-cenoses on Stratiotes

plants in Dutch broads. – Verh. R.I.N. 11: 1-86.
Higler, L.W.G., F.F. Repko & A.G. Klink, 1982. De macrofauna, in het bijzonder Chironomidae, van de Renkumse beek en enige andere sprengen. – Ongepubl. rapport, Leersum. 4 pp. + tabel.
Hildrew, A.G., C.R. Townsend & A. Hasham, 1985. The predatory Chironomidae of an iron-rich stream: feeding ecology and food web structure. – Ecol. Ent. 10: 403-413.
Holzer, M., 1980. Die Belebung der Gewässer von Sandkiesanschwemmungen unterhalb des aktiven Stromes des Flusses March in der Obermährischen Talsenkung. – Acta Univ. Palackianae Olomuc. Fac. Rerum Nat. 67: 107-129 (in Czech).
ICZN (1999). International Code of Zoological Nomenclature. Fourth edition. – International Trust for Zoological Nomenclature, London. 306 + XXIX pp.
Int Panis, L., B. Goddeeris & R.F. Verheyen, 1995. On the relationship between the oxygen microstratification in a pond and the spatial distributuion of the benthic chironomid fauna. – In: Cranston, P.S. (ed.): Chironomids. From genes to ecosystems. CSIRO Publ., East Melbourne, Australia: 323-328.
Izvekova, E.I., 1980. Pitanie. – Trudy vses. gidrobiol. obshch., zool. inst. AN SSSR 23: 72-101.
Janecek, B.F.U., 1995. Tanytarsus niger Andersen (Diptera: Chironomidae) and the chironomid community in Gebhartsteich, a carp pond in northern Austria. – In: Cranston, P. (ed.): Chironomids, from genes to ecosystems. CSIRO Publications, East Melbourne: 281-296.
Johnson, C.G., 1969. Migration and dispersal of insects by flight. – London, Methuen. XXII + 763 pp.
Jónasson, P.M., 1972. Ecology and production of the profundal benthos in relation to phytoplankton in lake Esrom. – Oikos Suppl. 14: 1-148.
Jónasson, P.M., 1977. Lake Esrom research 1867 – 1977. – Folia Limnol. Scand. 17: 67-89.
Kajak, Z., 1980. Role of invertebrate predators (mainly Procladius sp.) in benthos. – In: Murray, D.A. (ed.): Chironomidae. Ecology, systematics, cytology and physiology. Oxford, Pergamon Press: 339-348.
Kajak, Z. & K. Dusoge, 1970. Production efficiency of Procladius choreus Mg. (Chir. Dipt.) and its dependence of the trophic conditions. – Pol. Arch. Hydrobiol. 17 (30): 217-224.
Kajak, Z., K. Dusoge & A. Stanczykowska, 1968. Influence of mutual relations of organisms, especially Chironomidae, in natural benthic communities, on their abundance. – Ann. Zool Fenn. 5: 49-56.
Kalugina, N.S., 1959. O nekotorykh vozrastnykh izmeneniyakh v stroenii i biologii lichinok chironomid (Diptera Chironomidae). – Trudy vses. gidrobiol. obshch. Akad. Nauk SSSR 9: 85-107.
Kalugina, N.S., 1960. Die ontogenetischen Veränderungen in der Morphologie der Chironomidenlarven. – Verh. XI. Internat. Kongr. Ent. 1: 182-184.
Kawecka, B. & A. Kownacki, 1974. Food conditions of Chironomidae in the river Raba. – Ent. Tidsk. 95 Suppl.: 120-128.
Ketelaars, H.A.M., 1986. Makrofauna gemeenschappen in droogvallende watergangen. – Basisrapp. proj. EKOO 18: 1-111.. Prov. Waterstaat Overijssel, Zwolle.
Ketelaars, H.A.M., A.M.J.P. Kuijpers & L.W.C.A. van Breemen, 1993. Temporal and spatial distribution of chironomid larvae and oligochaetes in two Dutch storage reservoirs. – Neth. J. Aq. Ecol. 26: 361-369.
Klink, A., 1982. Rheopelopia ornata (Meigen): Description of the metamorphosis and ecology of a river inhabiting Tanypodinae-larva, new to the Dutch fauna (Diptera: Chironomidae). – Ent. Ber. 42: 78-80.
Klink, A., 1982a. Het genus Micropsectra Kieffer, een tax-

onomische en oekologische studie. – Medeklinker 2: 1-58, figs. 1-22. Wageningen.
Klink, A., 1983. Key to the Dutch larvae of Paratanytarsus Thienemann & Bause with a note on the ecology and the phylogenetic relations. – Medeklinker 3: 1-36.
Klink, A.G., 1990. Drift van makro-evertebraten in de Maas. – Rijkswaterstaat dienst binnenwateren / RIZA, nota 90.071. 45 pp. + bijl.
Klink, A., 1991. Maas 1986 – 1990. Evaluatie van 5 jaar hydrobiologisch onderzoek van makro-evertebraten. – Rapp. Meded. Hydrobiol. Adviesbur. Klink 39. 38 pp. + bijl.
Klink, A.G. & H. Moller Pillot, 1982. Onderzoek aan de makro-evertebraten in de grote Nederlandse rivieren. – Wageningen/Tilburg (private publ.). 57 pp.
Klink, A. & H. Moller Pillot, 1996. Lijst van de Nederlandse Chironomidae. – Werkgroep Ecologisch Waterbeheer, Themanr. 08. Groningen. 10 + 7 pp.
Kobayashi, T., 1998. Seasonal changes in body size and male genital structures of Procladius choreus (Diptera: Chironomidae: Tanypodinae). – Aq. Insects 20: 165-172.
Kobayashi, T., 2000. Procladius of Japan (Ins., Dipt., Chir., Tanypodinae). – In: Hoffrichter, E.O. (ed.): Late 20th century research on Chironomidae. Shaker verlag, Aachen: 143-146.
Konstantinov, A.S., 1958. Biologiya Chironomid i ich razvedenie. – Trudy Sarat. otdel. VNIORCh 5: 1-362.
Konstantinov, A.S., 1961. Feeding in some predatory chironomid larvae. – Vopr. Ichthiol. 1 (3) 20: 570-582. (Russian with english summary).
Koreneva, T.A., 1959. Ob otkladke yaits samkami Pelopiinae (Diptera: Tendipedidae) v Uchinskom vodokhranilishche. – Trudy vses. gidrobiol. obshch. Akad. Nauk SSSR 9: 108-120.
Koreneva, T.A., 1960. Ecology and taxonomy of Pelopiinae in the Uchinsk water reservoir. II. Pelopia, Ablabesmyia, Clinotanypus (Diptera, Tendipedidae). – Ent. Obozr. 39: 134-143.
Koskenniemi, E. & P. Sevola, 1989. Winter regulation effects on littoral chironomids in Hungarian reservoirs. – Acta Biol. Debr. Oecol. Hung. 3: 215-218.
Kouwets, F.A.C. & C. Davids, 1984. The occurrence of chironomid imagines in an area near Utrecht (the Netherlands), and their relations to water mite larvae. – Arch. Hydrobiol. 99: 296-317.
Kownacki, A. & M. Kownacka, 1968. Die Larve des Nilotanypus dubius (Meigen) 1804 (Diptera, Chironomidae). – Acta Hydrobiol., Krakow 10: 343-347.
Krebs, B.P.M., 1981. Aquatische macrofauna van binnendijkse wateren in het Deltagebied. I. Zuid-Beveland. – Delta Inst. Hydrobiol. Onderz., Rapp. & Versl. 1981-8: 1-158.
Krebs, B.P.M., 1984. Aquatische macrofauna van binnendijkse wateren in het Deltagebied. II. Zeeuws-Vlaanderen, oostelijk deel. – Delta Inst. Hydrobiol. Onderz., Rapp. en Versl. 1984-2: 1-124.
Krebs, B.P.M., 1990. Aquatische macrofauna van binnendijkse wateren in het Deltagebied. IV. Schouwen-Duiveland. – Delta Inst. Hydrobiol. Onderz., Rapp. en Versl. 1990-07: 1-124.
Krebs, B.P.M. & H.K.M. Moller Pillot, in prep. Influence of some environmental factors on the abundance of Chironomidae in a predominantly brackish water area.
Krenke, G. Ya., 1968. Novye dannye o pitanii lichinok Anatopynia varia F. – Nauchn. dokl. vysshej shkolu, Biol. nauki 5: 9-10.
Kreuzer, R., 1940. Limnologisch-ökologische Untersuchungen an holsteinischen Kleingewässern. – Arch. Hydrobiol. Suppl. 10: 359-572.
Kuiper, R. & J. Gardeniers, 1998. Taxonomy and ecology

of two Macropelopia species. – Lezing Nederlandse Chironomidendag. wageningen (unpubl.).

Ladle, M., J.S. Welton & J.A.B. Bass, 1984. Larval growth and production of three species of Chironomidae from an experimental recirculating stream. – Arch. Hydrobiol. 102: 201-214.

Langton. P. H., 1984. A key to pupal exuviae of British Chironomidae. – P.H. Langton, March, Cambridgeshire (private publ.). 324 pp.

Langton, P.H., 1991. A key to pupal exuviae of West Palaearctic Chironomidae. – P.H. Langton, Huntingdon, Cambridgeshire (private publ.). 386 pp.

Larsen, J.E., 1992. The phenology of Procladius crassinervis (Zett.), Procladius signatus (Zett.) and Procladius choreus (Mg.) (Dipt. Chir.) at Lake Hald, Denmark. – Neth. J. Aquat. Ecol. 26: 293-295.

Laville, H., 1971. Recherches sur les Chironomides (Diptera) lacustres du massif de Néouvielle (Hautes-Pyrénées). – Ann. Limnol. 7: 173-332.

Laville, H. & N. Giani, 1974. Phénologie et cycles biologiques des Chironomides de la zone littorale (0-7 m) du lac de Port-Bielh (Pyrénées centrales). – Ent. Tidskr. Suppl. 95: 139-155.

Laville, H. & G. Vinçon, 1991. A typological study of Pyrenean streams: comparative analysis of the Chironomidae (Diptera) communities in the Ossau and Aure valleys. – Verh. Intern. Verein. Limnol. 24: 1775-1784.

Learner, M.A. & D.W.B. Potter, 1974. The seasonal periodicity of emergence of insects from two ponds in Hertforshire, England, with special reference to the Chironomidae (Diptera: Nematocera). – Hydrobiologia 44: 495-510.

Leathers, A.L., 1922. Ecological study of aquatic midges and some related insects with special reference to feeding habits. – Bull. Bur. Fish., Wash. 1921-1922, 38: 1-61.

Lehmann, J., 1971. Die Chironomiden der Fulda. – Arch. Hydrobiol. Suppl. 37: 466-555.

Lellák, J., 1968. Positive Phototaxis der Chironomiden-Larvulae als regulierender Faktor ihrer Verteilung in stehenden Gewässern. – Ann. Zool. Fenn. 5: 84-87.

Lempke, B.J., 1962. Insecten gevangen op het lichtschip "Noord Hinder". – Ent. Ber., Amsterdam 22: 101-112.

Lenz, F., 1936. Tendipedidae (Chironomidae). a) Subfamilie Pelopiinae (Tanypodinae). B. Die Metamorphose der Pelopiinae. – In: Lindner, E. (ed.): Die Fliegen der palaearktischen Region 13b: 51-81.

LeSage, L. & A.D. Harrison, 1980. The biology of Cricotopus (Chironomidae, Orthocladiinae) in an algal-enriched stream: Part I. Normal biology. – Arch. Hydrobiol. Suppl. 57: 375-418.

Leuven, R.S.E.W., J.A. van der Velden, J.A.M. Vanhemelrijk & G. van der Velde, 1987. Impact of acidification on chironomid communities in poorly buffered waters in the Netherlands. – Ent. scand. Suppl. 29: 269-280.

Lindegaard, C., 1992. Zoobenthos ecology of Thingvallavatn: vertical distribution, abundance, population dynamics and production. – Oikos 64: 257-304.

Lindegaard, C., 1995. Chironomidae (Diptera) of European cold springs and factors influencing their distribution. – J. Kansas Ent. Soc. 68 (2) suppl.: 108-131.

Lindegaard, C., 1997. Diptera Chironomidae, Non-biting Midges. – In: Nilsson, A.: Aquatic Insects of Northern Europe. Apollo Books, Stenstrup (Denmark): 265-294.

Lindegaard, C. & K.P. Brodersen, 2000. The influence of temperature on emergence periods of Chironomidae (Diptera) from a shallow Danish lake. - In: Hoffrichter, E.O. (ed.): Late 20th century research on Chironomidae. Shaker verlag, Aachen: 313-324.

Lindegaard, C. & P.M. Jónasson, 1975. Life cycles of Chironomus hyperboreus Staeger and Tanytarsus graci-

lentus (Holmgren) (Chironomidae, Diptera) in Lake Myvatn, Northern Iceland. – Verh. Intern. Verein. Limnol. 19: 3155-3163.

Lindegaard, C. & E. Jónsson, 1987. Abundance, population dynamics and high production of Chironomidae (Diptera) in Hjarbaek Fjord, Denmark, during a period of eutrophication. – Ent. Scand. Suppl. 29: 293-302.

Lindegaard, C. & E. Mortensen, 1988. Abundance, life history and production of Chironomidae (Diptera) in a danish lowland stream. – Arch. Hydrobiol. Suppl. 81: 563-587.

Lindegaard, C., J. Thorup & M. Bahn, 1975. The invertebrate fauna of the moss carpet in the Danish spring Ravnkilde and its seasonal, vertical and horizontal distribution. – Arch. Hydrobiol. 75: 109-139.

Lindegaard-Petersen, C., 1972. An ecological investigation of the Chironomidae (Diptera) from a Danish lowland stream (Linding Å). Arch. Hydrobiol. 69: 465-507.

Lloyd, L., 1943. Materials for a study in animal competition. – Annls appl. Biol. 30: 47-60, 358-364.

Loden, M.S., 1974. Predation by chironomid (Diptera) larvae on oligochaetes. – Limnol. Oceanogr. 19: 156-159.

Mackey, A.P., 1976. Quantitative studies on the Chironomidae (Diptera) of the rivers Thames and Kennet. I. The Acorus zone. – Arch. Hydrobiol. 78: 240-267.

Mackey, A.P., 1977. Growth and development of larval Chironomidae. – Oikos 28: 270-275.

Mackey, A.P., 1979. Trophic dependencies of some larval Chironomidae (Diptera) and fish species in the River Thames. – Hydrobiologia 62: 241-247.

Markošová, R., 1979. Development of the periphytic community on artificial substrates in fish ponds. – Internat. Rev. ges. Hydrobiol. 64: 811-825.

Marlier, G., 1951. Le Smohain - La biologie d'un ruisseau de plaine. – Kon. Belg. Inst. Natuurwetensch., Verh. 114: 1-98.

Mason, C.F. & R.J. Bryant, 1975. Periphyton production and grazing by chironomids in Alderfen Broad, Norfolk. – Freshw. Biol. 5: 271-277.

Matena, J., 1989. Seasonal dynamics of a Chironomus plumosus (L.) (Diptera, Chironomidae) population from a fish pond in southern Bohemia. - Internat. Rev. ges. Hydrobiol. 74: 599-610.

Matena, J, 1990. Succession of Chironomus Meigen species (Diptera, Chironomidae) in newly filled ponds. – Internat. Rev. ges. Hydrobiol. 75: 45-57.

McCauley, V.J.E., 1974. Instar differentiation in larval Chironomidae (Diptera). – Can. Ent. 106: 179-200.

McLachlan, A.J., 1983. Life-history tactics of rain-pool dwellers. – J. Animal Ecol. 52: 545-561.

McLachlan, A.J., 1985. The relationship between habitat predictability and wing length in midges (Chironomidae). – Oikos 44: 391-397.

Michiels, S., 2004. Die Zuckmücken (Diptera: Chironomidae) der Elz – ein Beitrag zur Limnofauna des Schwarzwaldes. – Mitt. bad. Landesver. Naturkunde u. Naturschutz, N.F. 3: 111-128.

Michiels, S. & M. Spies, 2002. Description of Conchapelopia hittmairorum, spec. nov., and redefinition of similar western Palaearctic species. – Spixiana 25: 251-272.

Minshall, G.W., 1984. Aquatic insect-substratum relationships. – In: Resh, V. & D.M. Rosenberg (ed.): The ecology of aquatic insects. Praeger, New York: 358-400.

Moller Pillot, H.K.M., 1971. Faunistische beoordeling van de verontreiniging in laaglandbeken. – Tilburg. 285 pp.

Moller Pillot, H.K.M., 1984. De larven der Nederlandse Chironomidae Diptera). 1 A: Inleiding, Tanypodinae en Chironomini. – St. E.I.S. Nederland, Leiden. 277 pp.

Moller Pillot, H., 2003. Hoe waterdieren zich handhaven in een dynamische wereld. – Stichting Het

Noordbrabants Landschap, Haaren. 182 pp.

Moller Pillot, H.K.M. & P.L.T. Beuk, 2002. Family Chironomidae. – In: Beuk, P.L.T. (ed.) Checklist of the Diptera of the Netherlands. KNNV, Utrecht: 109-128.

Moller Pillot, H.K.M. & R.F.M. Buskens, 1990. De larven der Nederlandse Chironomidae (Diptera). 1 C: Autoekologie en verspreiding. – St. E.I.S. Nederland, Leiden. 87 pp.

Moller Pillot, H. & B. Krebs, 1981. Concept van een overzicht van de oekologie van chironomidelarven in Nederland. – Stencil, s.l. 41 pp.

Moller Pillot, H. & M. Schreijer, 1980. Natuur en landschap in Goirle en de Hilver. 1. De waterlopen. – Rapp. SBB Tilburg. 53 pp. + map.

Moog, O. (ed.), 1995. Fauna aquatica Austriaca. Katalog zur autökologischen Einstufung aquatischer Organismen Österreichs. – Wien, Bundesministerium Land- und Forstwirtschaft, loose-leaf.

Mossberg, P. & P. Nyberg, 1980. Bottom fauna of small acid forest lakes. – Report Institute Freshw. Res., Drottningholm 58: 77-87.

Mozley, S.C., 1979. Neglected characters in larval morphology as tools in taxonomy and phylogeny of chironomidae (Diptera). – Ent. scand. Suppl. 10: 27-36.

Mundie, J.H., 1957. The ecology of Chironomidae in storage reservoirs. – Trans. R. ent. Soc. Lond. 109: 149-232.

Mundie, J.H., 1965. The activity of benthic insects in the water mass of lakes. – Proc. 12th Internat. Congr. Ent. London: 410.

Munsterhjelm, G., 1920. Om Chironomidernas Ägglägging och Äggrupper. – Acta Soc. Fauna Flora Fenn. 47: 1-174.

Muragina-Koreneva, T.A., 1957. The ecology and systematics of the Pelopiinae (Dipt., Tendipedidae) of the Utsha reservoir, vicinity of Moscow. – Ent. Obozr. 36: 436-450. (in Russian)

Murray, D.A., 1976. Thienemannimyia pseudocarnea n. sp., a palaearctic species of the Tanypodinae (Diptera: Chironomidae). – Ent. scand. 7: 191-194.

Murray, D.A. & E.J. Fittkau, 1989. The adult males of Tanypodinae (Diptera: Chironomidae) of the Holarctic region Keys and diagnoses. – Ent. scand. Suppl. 34: 37-123.

Nijboer, R. & P. Verdonschot (red.), 2001. Zeldzaamheid van de macrofauna van de Nederlandse binnenwateren. – Werkgr. Ecol. Waterbeheer, themanr. 19: 1-77. (Basal data unpublished).

Nolte, U., 1993. Egg masses of Chironomidae (Diptera). A review, including new observations and a preliminary key. – Ent. scand. Suppl. 43: 1-75.

Olafsson, J.S., 1992. Vertical microdistribution of benthic chironomid larvae within a selection of the littoral zone of a lake. – Neth. J. Aq. Ecol. 26: 397-403.

Oliver, D.R., 1971. Life history of the Chironomidae. – Ann. Rev. Ent. 16: 211-230.

Orendt, C., 1993. Vergleichende Untersuchungen zur Ökologie litoraler, benthischer Chironomidae und anderer Diptera (Ceratopogonidae, Chaoboridae) in Seen des nördlichen Alpenvorlandes. – Thesis München Univ. 315 pp.

Orendt, C., 1999. Chironomids as bioindicators in acidified streams: a contribution to the acidity tolerance of chironomid species, with a classification in sensitivity classes. – Internat. Rev. Hydrobiol. 84: 439-449.

Orendt, C., 2002. Biozönotische Klassifizierung naturnaher Flussabschnitte des nördlichen Alpenvorlandes auf der Grundlage der Zuckmücken-Lebensgemeinschaften (Diptera: Chironomidae). – Lauterbornia 44: 121-146.

Palavesam, A. & J. Muthukrishnan, 1992. Influence of food quality and temperature on fecundity of Kiefferulus barbitarsis (Kieffer) (Diptera: Chironomidae). – Aq. Insects 14: 145-152.

Pankratova, V.Ya., 1970. Lichinki i kukolki komarov podsemejstva Orthocladiinae fauny SSSR (Diptera, Chironomidae = Tendipedidae). – Opredel. po faune SSSR 102: 1-343.

Pankratova, V.Ya., 1977. Lichinki i kukolki komarov podsemejstv Podonominae i Tanypodinae fauny SSSR (Diptera, Chironomidae = Tendipedidae). – Opredel. po faune SSSR 112: 1-153.

Parma, S. & B.P.M. Krebs, 1977. The distribution of chironomid larvae in relation to chloride concentration in a brackish water region of the Netherlands. – Hydrobiologia 52: 117-126.

Peters, A.J.G.P., R. Gijlstra & J.J.P. Gardeniers, 1988. Waterkwaliteitsbeoordeling van genormaliseerde beken met behulp van macrofauna. – STORA-rapport 88-06: 1-56 + bijl.

Pinder, L.C.V., 1978. A key to the adult males of the British Chironomidae (Diptera) the non-biting midges. – Freshw. Biol. Ass. Sc. Publ. 37: 1-169 + 189 figs.

Pinder, L.C.V., 1980. Spatial distribution of Chironomidae in an English chalk stream. – In: Murray, D.A. (ed.): Chironomidae: Ecology, Systematics, Cytology and Physiology. Oxford, Pergamon Press: 153-161.

Pinder, L.C.V., 1983. Observations on the life-cycles of some Chironomidae in southern England. – Mem. Amer. Ent. Soc. 34: 249-265.

Pinder, L.C.V., 1986. Biology of freshwater Chironomidae. – Ann. Rev.Ent. 31: 1-23.

Pinder, L.C.V., 1989. The adult males of Chironomidae (Diptera) of the Holarctic region – Introduction. – Ent. scand. Suppl. 34: 5-9.

Pinder, L.C.V., M. Ladle, T. Gledhill, J.A.B. Bass & A.M. Matthews, 1987. Biological surveillance of water quality – 1. A comparison of macroinvertebrate surveillance methods in relation to assessment of water quality, in a chalk stream. – Arch. Hydrobiol. 109: 207-226.

Raddum, G.G. & O.A. Saether, 1981. Chironomid communities in Norwegian lakes with different degrees of acidification. – Verh. Internat. Ver. theor. angew. Limnol. 21: 399-405.

Rasmussen, K. & C. Lindegaard, 1988. Effects of iron compounds on macroinvertebrate communities in a Danish lowland river system. – Water Res. 22: 1101-1108.

Reiss, F., 1968. Ökologische und systematische Untersuchungen an Chironomiden des Bodensees. Ein Beitrag zur lakustrischen Chironomidenfauna des nördlichen Alpenvorlandes. – Arch. Hydrobiol. 64: 176-323.

Reiss, F., 1971. Zum Kopulations-Mechanismus bei Chironomiden (Diptera) II. – Limnologica (Berlin) 8: 35-42.

Reiss, F., 1984. Die Chironomidenfauna (Diptera, Insecta) des Osterseengebietes in Oberbayern. – Ber. Akad. Naturschutz u. Landschaftspflege 8: 186-194.

Rieradevall, M. & S.J. Brooks, 2001. An identification guide to subfossil Tanypodinae larvae (Insecta: Diptera: Chironomidae) based on cephalic setation. – In: Journal of Palaeolimnology, 25: 81-99.

Rietveld, W., 1985. Typologie van sloten, een evaluatie. – Werkgr. Biol. Waterbeoordeling, subwerkgr. Typologie van sloten. Beekbergen. 17 pp. + bijl.

Ringelberg, J., 1976. Aquatische oecologie in het bijzonder van het zoete water. – Bohn, Scheltema & Holkema, Utrecht. 240 pp.

Roback, S.S., 1969. The immature stages of the genus Tanypus Meigen (Diptera: Chironomidae: Tanypodinae). – Trans. Am. Ent. Soc. 94: 407-428.

Roback, S.S., 1969a. Notes on the food of Tanypodinae larvae. – Ent. News 80: 13-18.

Roback, S.S., 1974. Insects (Arthropoda: Insecta). – In: Hart, C.W. & S.L.H. Fuller (eds.): Pollution ecology of freshwater invertebrates. Academic press, New York, London: 313-376.

Rodova, R.A., 1966. Razvitie *Cricotopus sylvestris* (Diptera, Chironomidae). – Trudy Inst. Biol. vnutr. Vod 12: 199-213.

Romaniszyn, W., 1950. Seasonal variations in the qualitative and quantitative distribution of the chironomid larvae in the Charzykowo Lake. – Pr. badaw. Inst. badaw. Lesn. 1: 99-157. (Polish).

Ruse, L., 2002. Chironomid pupal exuviae as indicators of lake status. – Arch. Hydrobiol. 153: 367-390.

Saether, O.A., 1977. Female genitalia in Chironomidae and other Nematocera: morphology, phylogenies, keys. – Bull. Fish. Res. Bd Can. 197: 209 pp.

Saether, O.A., 1979. Chironomid communities as water quality indicators. – Holarctic Ecol. 2: 65-74.

Saether, O.A., 1980. Glossary of chironomid morphology terminology (Diptera: Chironomidae). – Ent. scand. Suppl. 14: 1-51.

Samietz, R., 1999. Chironomidae.- Studia dipterologica Suppl. 2: 39-50. AMPYX-Verlag, Halle.

Schlee, D., 1968. Vergleichende Merkmalsanalyse zur Morphologie und Phylogenie der Corynoneura-Gruppe (Diptera: Chironomidae). Zugleich eine allgemeine Morphologie der Chironomiden-Imago . – Stuttg. Beitr. Naturk. 180: 1-150.

Schlee, D., 1977. Chironomidae als Beute von Dolichopodidae, Muscidae, Ephydridae, Anthomyidae, Scatophagidae und anderen Insecta. – Stuttg. Beitr. Naturk., A (Biologie) 302: 1-22.

Schleuter, A., 1985. Untersuchung der Makroinvertebratenfauna stehender Kleingewässer des Naturparkes Kottenforst-Ville unter besonderer Berücksichtigung der Chironomidae. – Thesis Bonn Univ. 217 pp.

Schmale, J.C., 1999. Hydrobiologisch onderzoek Berkheide 1994 – 1995 – 1996 – 1997. – N.V. Duinwaterbedrijf Zuid-Holland. 73 pp. + 80 bijl.

Schnabel, S., 1999. Faunistisch-ökologische Untersuchung der Chironomidae (Diptera: Nematocera) temporärer Tümpel in der Lahnaue bei Marburg. – Diplomarbeit Philipps-Univ. Marburg. 221 pp.

Schroevers, P. (red.), 1977. Beoordeling van de waterkwaliteit op basis van het mikrofytenbestand. – In: de Lange, L. & M.A. de Ruiter (red.): Biologische waterbeoordeling. Delft, TNO: 26-89.

Sergeeva, I.V., 1998. Vidovaya differentsirovka lichinok *Ablabesmyia* gr. *monilis* (L.) (Diptera, Chironomidae) iz vodoemov razlichnikh regionov Rossii. – Probl. ent. Rossii 2: 118.

Sergeeva, I.V., 2000. Volga river Tanypodinae (Dipt.: Chir.: Tanypodinae). – In: Hoffrichter, O. (ed.): Late 20th century research on Chironomidae. Aachen: 221-229.

Sergeeva, I.V., 2004. A review of the midge genus *Ablabesmyia* Johannsen, 1905 (Diptera, Chironomidae) of the fauna of Russia. – Ent. Rev. 84: 1033-1035.

Serra-Tosio, B. & H. Laville, 1991. Liste annotée des Diptères Chironomidés de France continentale et de Corse. – Annls. Limnol. 27: 37-74.

Shilova, A.I., 1976. Chironomidy Rybinskogo vodochranilishcha. – Nauka, Leningrad. 252 pp.

Shilova, A.I. & N.I. Zelentsov, 1972. The influence of photoperiodism on diapause in Chironomidae. – Inf. Byull. Inst. Biol. vnutr. Vod 13: 37-42. (in Russian).

Sládecek, V., 1973. System of water quality from the biological point of view. – Arch. Hydrobiol. Beiheft 7: 1-218.

Smit, H., 1982. De Maas. Op weg naar biologische waterbeoordeling van grote rivieren. – Rapport LH Vakgr. Natuurbeheer 667: 1-100.

Smit, H., 1995. Macrozoobenthos in the enclosed Rhine-Meuse Delta. – Thesis Nijmegen Univ. 192 + XIII pp.

Smit, H. & J.J.P. Gardeniers, 1986. Hydrobiologisch onderzoek in de Maas. Een aanzet tot biologische monitoring van grote rivieren. – H_2O 19: 314-322.

Smit, H., G. van der Velde & S. Dirksen, 1996. Chironomid larval assemblages in the enclosed Rhine-Meuse Delta: spatio-temporal patterns in an exposure gradient on a tidal sandy flat. – Arch. Hydrobiol. 137: 487-510.

Sokolova, N. Yu., 1968. Über die Ökologie der Chironomiden im Utscha-Stausee. – Ann. zool. fenn. 5: 139-143.

Sokolova, N. Yu., 1971. Life cycles of chironomids in the Uchinskoye reservoir. – Limnologica (Berlin) 8: 151-155.

Soponis, A.R. & C.L. Russell, 1982. Identification of instars in some larval *Polypedilum (Polypedilum)* (Diptera: Chironomidae). – Hydrobiologia 94: 25-32.

Spänhoff, B., N. Kaschek & E.I. Meyer, 2004. Laboratory investigation on community composition, emergence patterns and biomass of wood-inhabiting Chironomidae (Diptera) from a sandy lowland stream in Central Europe (Germany). – Aq. Ecol. 38: 547-560.

Spies, M. & O.A. Saether, 2004. Notes and recommendations on taxonomy and nomenclature of Chironomidae (Diptera). – Zootaxa 752: 1-90.

Staeger, C., 1839. Systematisk fortegnelse over de i Danmark hidtil fundne Diptera. – Naturhist. Tidsskr. 2: 549-600.

Steinmann, M., 1999. Einflüsse der saisonalen Überflutung auf die Chironomidenbesiedlung (Dipt.) aquatischer und amphibischer Biotope des unteren Odertals. – Thesis Freie Univ. Berlin. Aachen (Shaker Verlag). 117 pp.

Steenbergen, H.A., 1993. Macrofauna-atlas van Noord-Holland: verspreidingskaarten en responsies op milieufactoren van ongewervelde waterdieren. – Prov. Noord-Holland, Dienst Ruimte en Groen. Haarlem. 650 pp.

Sterba, O. & M. Holzer, 1977. Fauna der interstitiellen Gewässer der Sandkiessedimente unter der aktiven Strömung. – Vestnik Ceskosl. spolecn. zool. 41: 144-159.

?terba, O. & M. Holzer, 1977. Fauna der interstitiellen Gewässer der Sandkiessedimente unter der aktiven Strömung. – Vestnik Ceskosl. spolecn. zool. 41: 144-159.

Storey, A.W. & L.C.V. Pinder, 1985. Mesh-size and efficiency of sampling larval Chironomidae. – Hydrobiologia 124: 193-197.

Storey, A.W., 1987. Influence of temperature and food quality on the life history of an epiphytic chironomid. – Ent. scand. Suppl. 29: 339-347.

Strenzke, K., 1960. Die systematische und ökologische Differenzierung der Gattung Chironomus. – Ann. Ent. Fenn. 26: 111-138.

Tait Bowman, C.M., 1980. Emergence of chironomids from Rostherne Mere, England. – In: Murray, D.A. (ed.): Chironomidae. Ecology, systematics, cytology and physiology. Oxford, Pergamon Press: 291-295.

Tarwid, M. 1969. Analysis of the contents of the alimentary tract of predatory Pelopiinae larvae (Chironomidae). – Ekol. Polska A 17: 125-131.

Tauber, M.J. & C.A. Tauber, 1976. Insect seasonality: diapause maintenance, termination and postdiapause development. – Ann. Rev. Ent. 21: 81-107.

Tauber, M.J., C.A. Tauber & S. Masaki, 1986. Seasonal adaptations of insects. – Oxford Univ. Press.

Ten Winkel, E.H., 1987. Chironomid larvae and their foodweb relations in the littoral zone of lake Maarsseveen. – Thesis Amsterdam Univ. 145 pp.

Thienemann, A. & J. Zavrel, 1916. Die Metamorphose der Tanypinen. – Arch.Hydrobiol. Suppl. 2: 566-654.

Thienemann, A., 1954. *Chironomus.* Leben, Verbreitung und wirtschaftliche Bedeutung der Chironomiden. –

Binnengewässer 20: 1-834.

Timmermans, K.R., 1991. Trace metal ecotoxicokinetics of chironomids. – Thesis Amsterdam Univ. 185 pp.

Titmus, G. & R.M. Badcock, 1981. Distribution and feeding of larval Chironomidae in a gravel-pit lake. – Freshw. Biol. 11: 263-271.

Tokeshi, M., 1991. On the feeding habits of *Thienemannimyia festiva* (Diptera: Chironomidae). – Aq. Insects 13: 9-16.

Tokeshi, M., 1993. The structure of diversity in an epiphytic chironomid community. – Neth. J. Aq. Ecol. 26: 461-470.

Tolkamp, H. H., 1980. Organism-substrate relationships in lowland streams. – Thesis Wageningen. 211 pp.

Tõlp, Õ., 1971. Chironomid larvae in the brackish waters of Estonia. – Limnologica, Berlin 8: 95-97.

Tourenq, J.N., 1975. Recherches écologiques sur les Chironomides (Diptera) de Camargue. -Thesis Toulouse. 124 pp.

Tramper, N.M., 1979. Veedrinkputten als instabiele aquatische oecosystemen. – Delta Inst. Hydrobiol. Onderz., Stud. versl. D2: 1-74.

van Ee, G., 2000. Is natuurbeheer in randen langs bollenvelden zinvol? – Rapport Prov. Noord-Holland. 83 pp. + bijl.

Velden, J.A. van der, E.M. van Dam & S.M. Wiersma, 1995. The Chironomidae (Diptera) of Lake Volkerak-Zoommeer (The Netherlands) after freshening. – Lauterbornia 21: 139-147.

Verberk, W.C.E.P., H.H. van Kleef, M. Dijkman, P. van Hoek, P. Spierenburg & H. Esselink, 2005. Seasonal changes on two different spatial scales: response of aquatic invertebrates to water body and microhabitat. – Insect Science 12: 263-280.

Verdonschot, P.F.M., 2000. Natuurlijke levensgemeenschappen van de Nederlandse binnenwateren deel 1, Bronnen. - Rapport EC-LNV, nr. AS-01. Alterra (Wageningen). 86 pp.

Verdonschot, P.F.M., 2000a. Natuurlijke levensgemeenschappen van de Nederlandse binnenwateren deel 2, Beken. - Rapport EC-LNV, nr. AS-02. Alterra (Wageningen). 128 pp.

Verdonschot, P.F.M., L.W.G. Higler, W.F. van der Hoek & J.G.M. Cuppen, 1992. A list of macroinvertebrates in Dutch water types: a first step towards an ecological classification of surface waters based on key factors. – Hydrobiol. Bull. 25: 241-259.

Verdonschot, P.F.M. & J. A. Schot, 1987. Macrofaunal community-types in helocrene springs. – Res. Inst. Nature Management, 1987. Ann. Report 1986. Arnhem, Leersum and Texel: 85-103.

Verstegen, M., 1985. De macrofauna – met name de chironomidelarven – van een twaalftal vennen in de gemeenten Boxtel, Oisterwijk en Moergestel. – Verslag Utrecht Univ. 65 pp. + bijl.

Vodopich, D.S. & B.C. Cowell, 1984. Interaction of factors governing the distribution of a predatory aquatic insect. – Ecology 65: 39-52.

Waajen, G.W.A.M., 1982. Hydrobiologie van veenputten in de Mariapeel en de Liesselse Peel. – LH Wageningen, sektie Hydrobiol., verslag 82-1. 67 pp.

Walshe, B. M., 1948. The oxygen requirements and thermal resistance of chironomid larvae from flowing and from still waters. – J. exp. Biol. 25: 35-44.

Ward, G.M. & K.W. Cummins, 1978. Life history and growth pattern of *Paratendipes albimanus* in a Michigan headwater stream. – Annls Ent. Soc. Am. 71: 272-284.

Warwick, W.F., 1989. Morphological deformities in larvae of *Procladius* Skuse (Diptera: Chironomidae) and their biomonitoring potential. – Can. J. Fish. Aquat. Sci. 46: 1255-1270.

Warwick, W.F., 1991. Indexing deformities in ligulae and antennae of Procladius larvae (Diptera: Chironomidae): application to contaminant-stressed environments. – Can. J. Fish. Aquat. Sci. 48: 1151-1166.

Werkgroep Hydrobiologie MEC Eindhoven, 1993. De Groote Peel als leefmilieu voor aquatische macrofauna. – M.E.C. Eindhoven, 99 pp. + bijl.

Werkgroep Standaardisatie Macro-invertebraten Methoden en Analyse, 1999. Handleiding bemonsteringsapparatuur aquatische macro-invertebraten. – Melick: Werkgr. Ecol. Waterbeheer, Themanr. 17. 94 pp.

Wesenberg-Lund, C., 1943. Biologie der Süsswasserinsekten. – Berlin, Wien: Springer. 682 pp.

Wiederholm, T. (ed.), 1983. Chironomidae of the Holarctic region. Keys and diagnoses. Part 1. Larvae. – Ent. scand. Suppl. 19: 1-457.

Williams, C.J., 1982. The drift of some chironomid egg masses (Diptera: Chironomidae). – Freshw. Biol. 12: 573-578.

Wilson, R.S., 1977. Chironomid pupal exuviae in the river Chew. – Freshw. Biol. 7: 9-17.

Wilson, R.S. & L.P. Ruse, 2005. A guide to the identification of genera of chironomid pupal exuviae occurring in Britain and Ireland and their use in monitoring lotic and lentic fresh waters. – Freshw. Biol. Ass., Special Publ. 13: 1-176.

Wilson, R.S. & S.E. Wilson, 1984. A survey of the distribution of Chironomidae (Diptera, Insecta) of the river Rhine by sampling pupal exuviae. – Hydrobiol. Bull. 18: 119-132.

Wülker, W. & P. Götz, 1968. Die Verwendung der Imaginalscheiben zur Bestimmung des Entwicklungszustandes von *Chironomus*-Larven (Dipt.). – Zeitschr. Morphol. Tiere 62: 363-388.

Zavrel, J., 1936. Tanypodinen-Larven und -Puppen aus Partenkirchen. – Arch. Hydrobiol. 30: 318-326.

Zavrel, J. & A. Thienemann, 1921. Die Metamorphose der Tanypinen. II. – Arch. Hydrobiol. Suppl. 2: 655-784.

Zinchenko, T.D., 1997. Ekologicheskaya kharakteristika khironomid. – In: Noskova, O.L. (ed.): Ekologicheskaya bezopasnost i ustojchivoe razvitie Samarskoj oblasti 3: Ekologicheskoe sostoyanie bassejna reki Chapaevka v usloviyakh antropogennogo vozdejstviya. Togliatti: 183-198.

ACKNOWLEDGEMENTS

We wish to thank Dr B. Goddeeris, Leuven, Belgium for making helpful comments and producing drawings of some *Tanypus*-species and *Psectrotanypus* varius, and Dr Peter Langton for the permission to use his drawings of the thoracic horns. We thank Hub Cuppen (Adviesbureau Cuppen, Eerbeek, Netherlands), Hans Hop and Johan Mulder (Waterschap Groot Salland, Zwolle, Netherlands), Koeman & Bijkerk (Haren, Netherlands), Ad Kuijpers (Aqualab, Werkendam, Netherlands) and Barend van Maanen (Waterschap Roer en Overmaas, Sittard, Netherlands) for sending us larvae and information. Susanne Michiels (Emmendingen, Germany) and Dr Martin Spies (München, Germany) sent us slides showing the results of rearing larvae of important species. We also thank the following colleagues for testing the key and making comments: Hans Hop and Johan Mulder (Waterschap Groot Salland, Zwolle, Netherlands), Joep de Koning (Hoogheemraadschap van Delfland, Delft, Netherlands), Francien Lambregts (Waterschap Brabantse Delta, Breda, Netherlands), Barend van Maanen (Waterschap Roer en Overmaas, Sittard, Netherlands), Jeroen Meeuse (Waterschap Hunze en Aa's, Veendam, Netherlands), Mieke Moeleker and Maria Sanabria, GWL, Boxtel, Netherlands), André van Nieuwenhuijzen (Koeman & Bijkerk, Haren, Netherlands), Myra Swarte (RIZA, Lelystad, Netherlands), Saskia Wiersma (Waterschap Hollandse Delta, Rotterdam, Netherlands) and Michiel Wilhelm and David Tempelman (Grontmij-Aquasense, Amsterdam, Netherlands). Derek Middleton corrected our English and copy-edited the text.

A

TANYPODINAE-INDEX TO SCIENTIFIC NAME

	5.1 key page	5.3 comments page	5.4 matrix ch. page	5.4 matrix f. page	7(A) ecology page	photographs page, nr
Ablabesmyia	35, 43	50			74	
longistyla	35	50	58, 60		76	
monilis	35	50	58, 60	65, 66	75	III; 15, 16
phatta	35	50	58, 60		77	
Anatopynia	31, 45	50	56	62	78	II; 8, VII; 39
Apsectrotanypus	33, 45	50	56	62	79	I; 3
Arctopelopia	41, 42	50	58, 60	64, 67, 68, 69	81	VI; 33, 34, 35
Clinotanypus	29, 45	51	56		82	VII; 40
Conchapelopia	37	51			84	
hittmairorum	46		58		87	
melanops	42, 46		58, 60	64, 67, 68, 69	84	V; 32, VI; 36
pallidula	46			67, 68	87	
Guttipelopia	35, 43	51	58, 60	65	87	III; 14
Krenopelopia	39, 47	51	58, 60	64, 66	89	IV; 24
Labrundinia	35, 46	51	58, 60	65	90	
Larsia	48					
Macropelopia	33	52	56		90	I; 3, III; 13
adaucta	33, 45		56	63	91	
nebulosa	33, 45		56	62	92	
notata	33, 45		56		94	
Monopelopia	37, 47	52	58, 60	65, 66	94	
Natarsia	37, 47	52	58, 60	64, 66	96	I; 6, III; 18
Nilotanypus	35, 48	52	60		98	
Paramerina		52				
cingulata	39, 41, 48	52	58, 60	65, 66	99	V; 26
divisa	48	52			100	

COLOPHON

Authors
Henk J. Vallenduuk
Henk K.M. Moller Pillot

Illustrations and photography
Henk J. Vallenduuk
Boudewijn Goddeeris and Peter Langton,
names stated under the figures

Graphic layout and design
Erik de Bruin, Varwig Design, Hengelo

Cover illustrations
Background illustration: Xenopelopia - claws
Front cover illustration: Procladius - Head and thorax of prepupa

This publication was financially supported by:
Stowa, Utrecht
Riza, Lelystad

© KNNV Publishing, Zeist, The Netherlands
2nd edition (POD) 2013
ISBN 978-90-5011-259-8.
NUR 432
www.knnvpublishing.nl